Informing Choices for Meeting China's Energy
Challenges

Zheng Li · Angelo Amorelli
Pei Liu

Informing Choices
for Meeting China's Energy
Challenges

 Springer

Zheng Li
Tsinghua University
Beijing
China

Pei Liu
Tsinghua University
Beijing
China

Angelo Amorelli
BP International Ltd.
Sunbury-on-Thames
UK

ISBN 978-981-10-9596-2   ISBN 978-981-10-2353-8   (eBook)
DOI 10.1007/978-981-10-2353-8

This Springer imprint is published by Springer Nature
The registered company is Springer Science+Business Media Singapore Pte Ltd.

Foreword

BP likes to work with the best universities around the world to research the world's biggest energy challenges. Our collaboration with Tsinghua University allows us to consider China's challenges while ensuring that academic independence is maintained. BP has provided support to the research in this publication but not directed its conclusions.

Like many countries, China's demand for energy is expected to grow significantly as a result of sustained economic growth. We expect energy supply in China to continue being dominated by fossil fuels through to 2050. Hence China will need to import an increasing proportion of its energy supplies, particularly oil and gas. The security and diversity of these supplies are of national importance.

There are also other natural resource constraints. Having access to supplies of fresh water, protecting air and water quality are significant future challenges. The macro-economic model that has successfully driven China's growth—which has been based on inexpensive land, early stage environmental regulation and the ability to employ migrant workers from across China to manufacture goods for export—is unlikely to be the preferred model into the next phase of development. To tackle these and many other challenges, central and provincial administrations have called for structural reform of the economy: upgrading traditional industries and developing advanced ones; further building of energy, transport, and IT infrastructure; and fostering innovation. Innovation is critical to China's ability to sustain its growth. We hope this publication can help to inform this future strategy.

David Eyton
BP Group
Head of Technology

Preface 1

We have been collaborating with BP for over 10 years. Our partnership combines Tsinghua's academic research strength with BP's international energy expertise, applied to some of China's biggest development challenges. As China's economy is going through a new era, it also brings new challenges to energy security, clean and low-carbon energy utilisation, and environmental protection. In the last 5 years, the collaboration between Tsinghua and BP has made steady progress towards systematic analysis of these challenges, based on sound engineering principles, across several topics: analysing the energy embodied in infrastructure and the inefficiencies of overcapacity building with shorter useful lifetime or less operating time; revealing the huge disparities in regional energy consumptions, and the relationship between technology choices and changing energy prices. Our improved understanding of these outcomes is expected to help better plan China's energy future.

Prof. Zheng Li
Director Tsinghua/BP Clean Energy Centre

Preface 2

China has become an economic superpower over the last 20 years and will continue to grow substantially for the foreseeable future. All nations require energy to fuel economic growth. It is increasingly obvious that China needs to find a more sustainable path to power this economic growth.

This publication has attempted to describe the scale and complexity of China's energy system. We have used Sankey diagrams to visualise the energy flows across the economy. We have also looked to visualise and categorise the differences between China's provinces. China's provinces extend over a vast geography and are at different stages of economic development. Regional planning and inter-regional optimisation will be essential for planning future infrastructure that will allow China as a whole to optimise and avoid overcapacity and inefficiency.

Consistent with this analysis, we have introduced our energy systems modelling approach and demonstrated its application to the power sector. Our power model can help to optimise the overall design of the power system and reveal how different future investment choices only emerge, e.g. natural gas-fired power, as the level of detail is increased, such as regional, seasonal, and diurnal demand factors are considered. We also begin to look at options for mitigating carbon and the major role that renewables and natural gas could play as well as carbon pricing.

Planning a more sustainable future is now critically important for China as there are major risks that need to be dealt with today. Three of the key challenges are as follows:

- The Chinese economy is exposed to overcapacity, including that of energy and associated infrastructure.
- Urban air quality regularly exceeds WHO recommended limits by up to an order of magnitude.
- Coal dominates the energy supply mix, with major environmental consequences for air and water quality, as well as the global climate.

There are also three major considerations:

- China is heterogeneous with energy-related issues varying by province. No single national policy will fit each province's economic and environmental needs.
- Renewables are growing rapidly in China, due to the policy emphasis on energy security, the environment, and the global climate but managing their intermittency needs to be modelled in great detail hour by hour, region by region.
- The growth in energy demand is slowing as China has industrialised and looks to upgrade its economy. Future growth will be policy-dependent, particularly related to China's commitments to carbon emissions reductions under the Paris Agreement.

This leads us to our major recommendations for prioritisation of strategic options to deal with the three challenges:

1. To deal with overcapacity:
   Planners must consider differences between provinces and consider inter-regional imports and exports of energy and the related infrastructure to develop a more efficient and cost-optimised configuration.

   There should be a focus on building energy infrastructure that prioritises quality over quantity, and that delivers long-term competitiveness over short-term GDP growth.

2. To deal with environmental challenges:
   China can learn much from North America and Europe in dealing with the environmental impact of industrial development. The technical solutions are available.

   The enforcement of environmental regulations and the use of clean-up technologies are essential.

   Reducing the use of energy through improvement of energy efficiency in industry, buildings, and transport should be prioritised.

   Distributed direct burning of raw coal with no pollution control measures should be replaced as a priority with clean natural gas or demand side electrification.

3. To diversify the energy supply mix to a more low-carbon future:
   China must accelerate the growth in indigenous unconventional shale gas supply, and that requires restructuring of the market, and provide open access to attract investors and participants with the appropriate expertise.

   China could then better prioritise the use of natural gas in power and balance intermittent renewable supply. Even without indigenous gas supply, the import of natural gas to provide cleaner power has significant environmental and operational merits.

   The path to a more low-carbon future will requires further exploitation of renewable energy and the development of advanced gas turbine and nuclear capabilities. Where possible, China should look to maximise the use of waste biomass locally for heat and power.

Although the share of coal in total energy mix will decrease, it will not disappear overnight, especially because many coal facilities such as coal power plants are brand-new and a lot of carbon and energy have been invested in their construction. While it is always important to ensure their operational efficiency, it is also imperative for coal power plants to back up renewable energy to optimise carbon emissions of the whole system.

Carbon capture and storage has many challenges and is far from being proven. Therefore relying on CCS to mitigate the long-term use of coal carries significant uncertainty. So it is necessary to continue to strengthen research on CCS.

Beijing, China   Zheng Li
Sunbury-on-Thames, UK   Angelo Amorelli
Beijing, China   Pei Liu

Acknowledgements

This work is based on a five-year collaborative project between BP and Tsinghua University to analyse energy flows between key energy supply, conversion, and consumption sectors in China out to 2050, and provide an integrated perspective regarding the macro-energy system of China.

Both BP and Tsinghua University would like to acknowledge the work of the academics, students, and BP staff who have contributed to this collaboration over the past 5 years. In particular, we would like to acknowledge Ma Linwei, Wang Zhe, Xue Yali, Xu Zhaofeng, and Li Weiqi from Tsinghua University, and Ian Jones, Anna-Marie Greenaway, and Bokun Qin from BP, for their valuable contribution to this book.

In addition we would like to recognise the support of our steering board which consists of renowned Tsinghua Academics, Profs. Chen Jining, Ni Weidou, and He Jiankun, and Energy Company leaders, David Eyton (Fellow of Royal Academy of Engineering) and Zhang Yuzhuo (Fellow of Chinese Academy of Engineering).

January 2016

Contents

Chapter 1
Introduction

China has invested and continued to invest in building out infrastructure across industrial, transport, and urban sectors. In 2013, China produced over half of the steel, cement, and flat glass in the world, and is the largest country in terms of energy production and consumption. As in any developing economy, there is a risk of inefficient investment in chasing growth targets due to the creation of over-capacity. For China, the problem is serious. There is huge overcapacity today in several industrial sectors such as steel, cement, chemicals, and power generation. The overcapacity manifests itself as inefficient assets, many locked into early gen-eration technologies, and some bringing serious environmental issues. Facing the future, how can China be smarter in its industrial investments by avoiding more overcapacity?

As China seeks to upgrade its economy, the country cannot afford to waste investment or energy in building new but under-utilised industrial infrastruc-ture. Otherwise the country could risk becoming economically less competitive. Meanwhile, China still needs to upgrade its infrastructure at a huge scale, particu-larly in the power and refining sectors to solve the country's severe environmen-tal challenges. This could create further inefficiency if not executed smartly with deeper assessment of the options.

In the energy sector, the investment risks are increasing rather than diminishing because the diversity of investment options has increased. Twenty years ago, there was coal or hydro power, now there are nuclear, wind, solar, and gas as well. Ten years ago there were only gasoline or diesel vehicles, now hybrids, electrical, bio-fuels, and natural gas (NG) vehicles are emerging to sit alongside with advanced internal combustion engines (ICE). More complicated investment decisions also have to be made with the backdrop of volatile oil prices and the changing rela-tionship between the government and State Owned Enterprises. How does China choose the right path?

© Springer Science+Business Media Singapore 2016
Z. Li et al., *Informing Choices for Meeting China's Energy Challenges*,
DOI 10.1007/978-981-10-2353-8_1

With increasing delegation to the local areas (such as provinces), there is also a need to consider China as a set of heterogeneous regions composed of different provinces with remarkably different characteristics, rather than a homogeneous country where one strategy fits all. Through recognising the differing characteristics and inter-play between provinces, it may be possible to pinpoint opportunities for optimisation of key networks such as mass transit, electricity generation and gas transportation. How does China optimise energy infrastructure across its vast geography?

Studies underway at the Tsinghua-BP Clean Energy Centre have been aimed at better understanding these multi-layered issues by modelling the multiplicity of choices for energy pathways and commercial value chains, in the power and transport sectors, and incorporating regional and environmental lenses.

This publication provides an initial analysis of China's current and future energy system's challenges, focused on the power sector. Our analysis is based on an understanding of the underlying technologies, essentially an understanding of engineering, and the use of detailed mathematical modelling to describe and illustrate the complex energy system and its interactions, under different scenarios.

Such modelling is totally dependent on the quality of the data available and also the robustness of the assumptions underlying potential scenarios.

This report will take the reader through the work carried out at the Tsinghua Clean Energy Centre in a systematic fashion:

Chapter 2 provides the context. In this chapter, we provide an overview of China's energy system and energy flows. We employ Sankey diagrams to graphically illustrate these flows. In addition we highlight the differences between China's provinces.

Chapter 3 discusses the challenges and uncertainties of long-term energy forecasting by reviewing past forecasts from the world's most respected and quoted sources.

Chapters 4, 5 consider China's Fossil Fuel Resources, the importance of coal and the untapped potential of unconventional gas and the overcapacity issues of the current energy infrastructure.

Chapter 6 summarises our view of the state of the art and potential of energy technologies, critical for underpinning any projections of the future.

Chapters 7, 8 introduce our modelling approach and its application to the power sector. Our power model can help to optimise the power system and reveal how different future investment choices emerge as the level of detail is increased, such as regional, seasonal and diurnal demand factors are considered. We also begin to look at mitigating carbon and the major role that NG and renewables could play as well as carbon pricing.

Chapter 9 introduces the dynamic between China's economic growth and the pressures it creates on water resources. It discusses potential pathways for both policy and technology to alleviate pressure.

Chapter 10 describes some future low carbon options for distributed combustion power plants such as plants with carbon capture or exploiting municipal solid wastes.

Chapter 11 concludes main findings in the previous chapters.

Our scope is broad and all research is never finished. Nevertheless this publication can hopefully provide the reader with an illustration of an improved, fact-based, and systematic approach to defining China's future energy choices.

Chapter 2
China's Energy System

In this chapter we provide an overview of China's energy system and energy flows. We employ Sankey diagrams to graphically illustrate these flows, and discuss the dynamics of China's energy system by analysing key influential factors, including economic growth, energy resources, environmental impacts, and energy management. In addition we highlight the differences between China's provinces using a model that considers not only primary energy flows, but also the energy embedded in products and services that flow between these regions.

2.1 Energy Allocation Sankey Diagrams

The Sankey Diagram is used as the main tool to analyse and illustrate the physical system of energy use in China. This approach was first used by the Irish engineer Riall Sankey in 1898 (Schmidt 2008) and now has become a popular graphical tool for mapping energy flows from the resource through transformations (for example refining crude oil to produce clean diesel) to final consumption. In a Sankey diagram, the quantity of energy is traced by arrows or lines, with the line width being proportional to an energy flow. This simple graphical approach allows easy identification of dominant energy flows (Ma et al. 2012a, b).

The following Sankey diagrams are termed as "Energy Allocation Sankey Diagrams" as they do not exclude any energy or material losses during the individual transformational processes. The Sankey diagram energy or mass balance can be based on the first law of thermodynamics, or the second law of thermodynamics, and expressed in Joules (J) or the mass balance and expressed in Tonnes or Kilogrammes (t or kg). The only difference is the measurement of the amount of each flow. The first law of thermodynamics is based on the measurement of energy content by low heating values (LHV), and the second law of thermodynamics is

© Springer Science+Business Media Singapore 2016
Z. Li et al., *Informing Choices for Meeting China's Energy Challenges*,
DOI 10.1007/978-981-10-2353-8_2

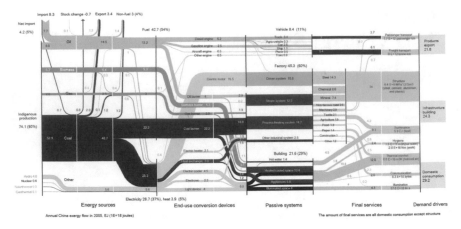

Fig. 2.1 2005 Energy Flows from Primary Sources to Demand Drivers of Chinese Society (Ma et al. 2012a)

based on the measurement of exergy[1] which can be derived from LHV, and the mass balance is based on the actual mass measurement of a fuel.

Using these methodologies, we have mapped the energy allocation Sankey diagram of China's total energy utilisation, coal flow, oil flow, and natural gas flow whose detailed mapping method and data source will be further introduced.

2.1.1 Total Energy Flows

Before looking at the latest energy data, our first Sankey diagram in Fig. 2.1 traces the total energy flows in China from a decade ago. It is an Energy Allocation Sankey Diagram for 2005, without showing any energy losses and based on exergy measurement. The method and data are explained in detail in a previous academic publication (Ma et al. 2012a). This diagram divides the technical process of energy utilisation into 5 stages, including Energy Sources, End-use Conversion Devices, Passive Systems, Final Services, and Demand Drivers, which were defined in the previous work (Ma et al. 2012a, b).

The Sankey Diagram underlines that the main features of China's energy system is the dominance of coal as a fuel and the large energy consumption in the heavy manufacturing sector (especially energy-intensive manufacturing industries such as Steel, Chemical, and Mineral). Looking closer at the data, we can see other major insights:

---

[1]In thermodynamics, the exergy of a system is the maximum useful work possible during a process that brings the system into equilibrium with its surroundings. Exergy analysis is performed in the field of industrial ecology to use energy more efficiently.

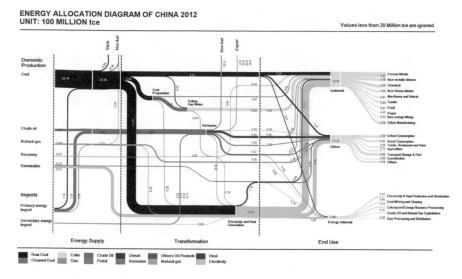

Fig. 2.2 2012 Energy Flows from Energy Supply to End-Use of Chinese Society (Chong et al. 2014)

Demand drivers: about two-third of energy consumption is driven by two major economic activities—product export (about one-third) and infrastructure building (about one-third). The latter is consistent with China's major investment in infrastructure building for economic development and the former reflecting China's rise as a global manufacturing powerhouse.

Final services: about 45 % of energy consumption is consumed to provide the final services for construction, e.g. energy-intensive products such as steel, cement, and plastic.

Passive systems: 60 % of energy is consumed by factories, consisting of intensive heat and electricity demand of steel, chemical, and mineral processing.

End-use conversion devices: the main devices consuming energy are coal burners (29 % of energy consumption) and electric motors (20 % of energy consumption). These mainly consume coal as direct fuel or consume electricity derived predominantly from coal power generation.

Energy sources: China's primary energy consumption is dominated by coal because it is cheap and abundant in China, and readily utilised to power economic growth and industrialisation (as it did in the West).

Figure 2.2 illustrates the updated energy flows of China in the year 2012, which is based on the measurement of LHV and only aggregates the process of energy use into 3 stages: energy supply, transformation, and end-use. The method and data have been published (Chong et al. 2014). Although the method differs from the previous work (Fig. 2.1), it also shows that China's energy system has not changed appreciably over the years; only the overall consumption has increased. Coal remains dominant (69.7 %) as the fuel of choice. Energy end-use in the economy is dominated by manufacturing (57 %).

2.1.2 Coal, Oil, and Natural Gas Flows

China's primary energy consumption, like all other major economies, is dominated by fossil fuel, especially coal. In this section we will further analyse the energy flows of coal, oil, and natural gas through the economy.

Figure 2.3 presents the coal flows of China in 2012 (Chong et al. 2014, 2015). It shows that the coal use in China is spread across multiple sectors, but mainly used for power/heat generation (almost 50 %), coking and direct fuel use. Four notable uses in the processing and manufacture of Ferrous metals (iron and steel), Non-metallic minerals (cement, glasses etc.), Chemicals, and Non-ferrous metals (aluminium, copper etc.), add up to circa 50 % of raw coal consumption, from their demand for electricity, heat, and coking products. Therefore, the past and future development of these energy-intensive manufacturing sectors shapes coal demand.

Figure 2.4 presents the oil flows of China in 2010 (the most recent robust data available) (Ma et al. 2012b) and references China's energy balance in 2010 (NBSC 2012). It shows the crude oil supply of China is heavily dependent on imports and crude oil is mainly used to produce transport fuels such as diesel and gasoline. Transportation shapes oil demand, and to a lesser extent feed stocks for the chemical industry.

Figure 2.5 presents the natural gas flow of China in 2012 (Geng 2015). It shows that the natural gas supply of China is also dependent on imports, just over 33.5 % of the total natural gas consumed in China. LNG imports are greater than pipeline imports in 2012. Though conventional natural gas production is still the main source of domestic gas production, unconventional natural gas has become an important source of gas production, 41.7 % of total natural gas produced in China. This locally produced gas includes tight gas (30 % of total production), coal bed methane (11.7 %), and very little shale gas. The natural gas supplied is mainly used as fuel for industry and building space heating (residential and commercial use).

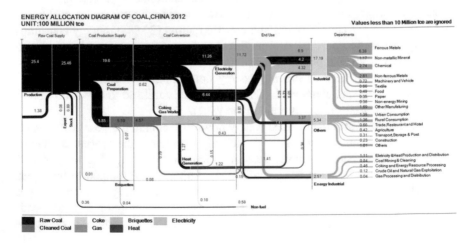

Fig. 2.3   2012 Coal flows of China

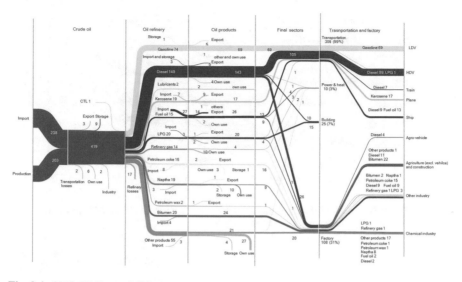

Fig. 2.4   2010 Oil flows of China

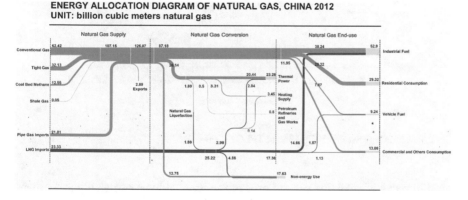

Fig. 2.5   2012 Natural gas flows of China

2.2   The Key Influencing Factors on China's Energy System

2.2.1   Economic Growth and Energy Demand

After 2000, China's economic growth entered a special stage in industrialization called "accelerated industrialization". Its main feature is the rapid expansion of heavy and chemical industry together with rapid urbanisation and motorisation. This economic growth comes with high energy intensity [energy consumption per unit of GDP (Gross Domestic Product)]. Figure 2.6 (Li et al. 2012) shows the

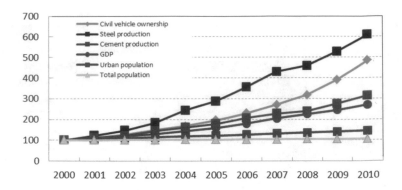

Fig. 2.6 Rapid urbanisation, motorization, and industrialization of China (all above curves are based on the data of 2000 as a relative index of 100)

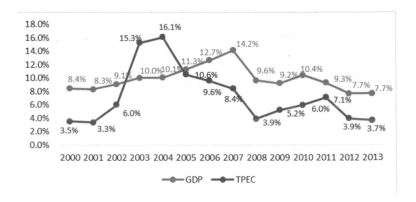

Fig. 2.7 China's growth rate of GDP and TPEC (2000–2013)

rapid increase of steel and cement production, civil vehicle ownership and urbanisation ratios over the first decade of this century.

Referring to Fig. 2.7, from 2000 to 2011, the average annual growth rates of GDP and TPEC (total primary energy consumption) in China were over 10 % and around 8.4 % respectively. However, after that in 2012 and 2013, the GDP growth rate decelerated to 7.7 %. Meanwhile, the TPEC growth rate of 2012 and 2013 fell below 4 % (NBSC 2014a, b).

The trend of lower GDP and TPEC growth is very likely to continue based on China's desire to restructure its economy away from heavy manufacturing; as signalled by 2013 data showing the proportion of tertiary industry (service industry) in GDP overpassed the proportion of secondary industry (manufacturing and construction) for the first time (NBSC 2013), and 2012 data indicates that final consumption expenditure has become the main driving force of GDP growth (NBSC 2014a).

In summary, China has entered a turning point. Politically this has been defined by the Chinese leadership as the "New Normal". This will mean that the future economic growth will be different and likely much less energy intensive than the past 15 years.

2.2.2   China's Energy Resources and Supply Structure

China is known to have relatively abundant coal resources and limited oil and natural gas resources. Hence it is unsurprising that the energy production in China is dominated by coal production (Fig. 2.8). The indigenous production of oil has been relatively stable since 2000. While coal production has steadily increased over the same period, China has witnessed the emergence of natural gas and renewable energy. Over the period from 2000 to 2013, natural gas production increased from 28.1 to 124.9 bcm (BP 2015).

Meanwhile, the rapid increase of non-fossil renewable and nuclear energy can be observed by the data of power generation capacity (CEC 2008–2013) as illustrated in Fig. 2.9. In 2013, the power generation capacity of non-fossil renewable and nuclear energy has occupied 31.6 % of the total power generation capacity (CEC 2014).

Even with the expansion of its indigenous energy production, China still cannot satisfy its own increasing energy demand, so there has been a steady increase of energy imports, including oil, natural gas and also notably high-grade coal. The phenomenon of coal imports arises from the costs of long distance transportation of coal across China from the producing Northern and Western Provinces to the consuming provinces of the South and East, which makes the import of high-grade coal more competitive in coastal regions.

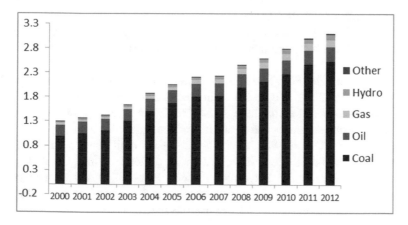

Fig. 2.8   China's primary energy production (2000–2012, unit in billion tce)

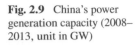 **Fig. 2.9** China's power generation capacity (2008–2013, unit in GW)

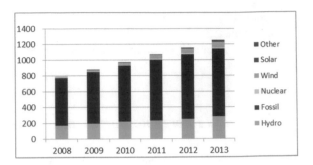

Over the period 2000–2013 (BP 2015), the import dependency of oil, natural gas, and coal has increased from 27.5 to 59 %, −11 to 27.6 %, and −1.9 to 4.4 % respectively.

2.2.3 China's Environmental Challenges

The scale of energy consumption (China is now the second largest economy in the world) and its reliance on coal, contribute to the emissions of major air pollutants and GHG gases. In 2012 (Xinhua Net 2014a, b), coal consumption, including direct coal combustion and activities directly linked with coal utilisation, contributed 93, 70, and 62 % of total emitted SO_2, NO_x, and primary PM 2.5 respectively. The contribution of energy-related CO_2 emissions to total CO_2 emission in China is around 90, and 80 % of energy-related CO_2 emissions came from coal utilisation.

To reduce emissions from energy generation, especially from coal combustion, China implemented much stricter emissions standards for air pollutants from thermal power generation (using coal, oil, and natural gas as fuel). The standard for coal power is listed in Table 2.1.

Moreover, China has endeavoured to reduce the proportion of coal in the total primary energy consumption and in 2013, the coal proportion slipped to 67.6 % (BP 2015).

Statistics from MEPC (Ministry of Environmental Protection of People's Republic of China 2003) (Fig. 2.10), indicated that the total emissions of SO_2 and NO_x from waste gas in China have started to fall in recent years, as a result of improved emission standards and treatment technologies.

Compared to NO_x and SO_2 air pollutants, CO_2 emissions from energy utilisation, especially coal combustion, are more difficult to mitigate and also at quantities several orders of magnitude greater. However, based on energy saving technologies and practices along with and the emergence of renewables, the increasing rate of energy-related CO_2 emissions in China has slowed.

Table 2.1 China's emissions standard of air pollutants from coal power plants (unit in mg/m³ flue gas)

Emission standard	Dust	SO_2	NO_x	Mercury and its compounds
GB 13223-2003 (old)[a]	50	400	450–1100[c]	No standard
GB 13223-2011 (new)[b]	30	100	100	0.03
For key regions	20	50	100	0.03

[a]The standard GB 13223-2003 applied for all projects of construction and retrofit of thermal power plants after Jan 1st, 2004. The data listed applied for most of regions and technologies except some special regions and technologies mentioned in it
[b]The standards GB 13223-2011 applied for all new construction projects of thermal power plants after Jan 1st, 2012 and further for all existing thermal power plants after July 1st, 2014, but the standard for mercury and its compounds only applied after Jan 1st, 2015. The data listed applied for most of regions and technologies except some special regions and technologies mentioned in it, while key regions mean some special regions need to control emissions more carefully
[c]The actual data (450, 650, or 1100) depends on the proportion of volatiles in coal, referring to GB 13223-2003

Fig. 2.10 China's Emissions of SO_2 and NOx in Waste Gas Emitted (unit in Mt) (in 2011 there was an adjustment of statistical method and scope, so the data is not so consistent with before)

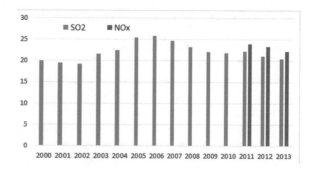

Figure 2.11 compares China to 4 other major countries, and China has delivered the most rapid reduction of the CO_2 intensity per unit of GDP in the period 1990–2012. In 2014, the total energy-related CO_2 emissions of China only increased by 0.9 % relative to 2013 (BP 2015).

2.2.4 China's Approach to Strategic Planning of Energy

To understand the strategic planning of energy development in China over the past 20 years, we first need to understand the political goals of the central government of China. The series of Presidential reports from 2000 to 2010 to the NCCPC (National Congress of Communist Party of People's Republic of China) (People's Net), contained several political goals.

Reports to the 15th and 16th NCCPC by President Jiang Zemin in 1997 and 2002 established the philosophy of 'opening and reforming' and the realisation of

Fig. 2.11 Trends in CO_2
Emission Intensities for the
Top Five Emitting Countries
(the size of the circle
represents the total CO_2
emissions from the country in
that year.)

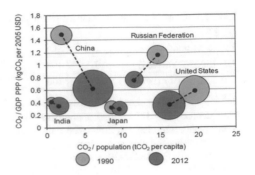

rapid economic growth. The key target was to double the GDP in 2010 relative to 2000 and further double it by 2020.

The report to the 17th NCCPC by President Hu Jintao in 2007 again emphasised the importance of rapid economic growth and sticking to the same GDP growth targets, but pointed out that economic growth must be based on the optimisation of economic structure, improvement of economic efficiency, reduction of resource consumption, and protection of environment. A new concept "scientific development" was proposed, which emphasised a more balanced development, not only single-minded economic growth.

Within the framework of these political goals, the main targets of the 10th (2001–2005) and 11th (2006–2010) Five-Year-Plans of economic and energy development are summarised as follows.

The 10th Five-Year-Plan: there was no clear target for energy development in the economic and social development plan, except an aim to reduce the emission of main pollutants by 10 %, which had indirect relationship with energy development. In the special plan of energy development, the emphasis was on energy security, but also mentioned the reduction of coal proportion in TPEC, the improvement of coal power generation efficiency and the reduction of energy intensity.

The 11th Five-Year-Plan: This announced a binding target of 20 % lower energy intensity for the first time in the economic and social development plan. In the special plan of energy development, it further proposed to reduce the proportion of oil consumption in addition to lower coal proportion in the TPEC, by the development of natural gas and non-fossil energy. The plan also emphasised the need to improve energy efficiency.

In the 12th Five-Year period (2011–2015), there were some changes both in political goals and targets:

The political goals announced in the report of President Hu Jintao to the 18th NCCPC: economic growth was still prioritised, as was the target to double GDP in 2020 compared to 2010. However, it was proposed 'to actively promote the development of an ecologic civilization' as one of the main missions in the report, to promote a 'revolution of energy production and consumption' to ensure energy security and protect the environment.

The 12th Five-Year-Plan also set a binding target of 16 % reduction of energy intensity of GDP, it was also proposed to reduce CO_2 intensity of GDP by 17 % and increase the proportion of non-fossil energy in the TPEC to 11.4 % in the economic and social development plan. It also clearly mentioned controlling the oil import dependency at 61 %, reducing the coal proportion in TPEC to 65 %, and increasing the natural gas proportion to 7.5 %, although these are only expected targets.

2.3   China's Regional Differences

China is too often defined internationally by its national statistics that hide the true heterogeneity of its geographical scale. In reality China consists of over 30 provinces and special regions, some the scale of major countries not only in area and population but also in economic output. It is critical to understand the differences between these regions to fully analyse China's current energy infrastructure and to optimise its future needs and investments. In this section, we begin to explore this in detail using the best available and consistent data, which comes from 2007.

The energy consumption of 30 regions defined by three different metrics is illustrated in Fig. 2.12 (Li et al. 2014). The 30 regions (column) are arranged from left to right by their per capita GDPs (PPP, current international $) of 2007. The main insights from this data are as follows:

(1) If we decompose energy consumption by Method 0, which only accounts flow-in and flow-out of energy products, the highest energy-consuming regions are segregated into two classes: those with huge energy (coal) exports (like Shanxi, Inner Mongolia, and Ningxia) and some of the most economically developed regions with large scale manufacturing (like Shanghai and Tianjin).

(2) If we decompose energy consumption by Method A, which is based on an input-output method and accounts energy embodied in goods and services for final consumption expenditure, gross capital formation and goods and services flow-out (export), we can discern that these high energy consumption regions normally have huge exports of embodied energy.

(3) If we decompose energy consumption by Method B, which only accounts embodied energy by goods and services flow-in (import) and domestic energy consumption, then the high energy consumption regions are the most highly developed (Beijing for example) and those with significant heavy industries (Liaoning for example).

What our early work on the regional disparities has revealed is how heterogeneous the provinces are. The differences are large. This must be considered in modelling the future and drafting future legislation–one size will not fit all.

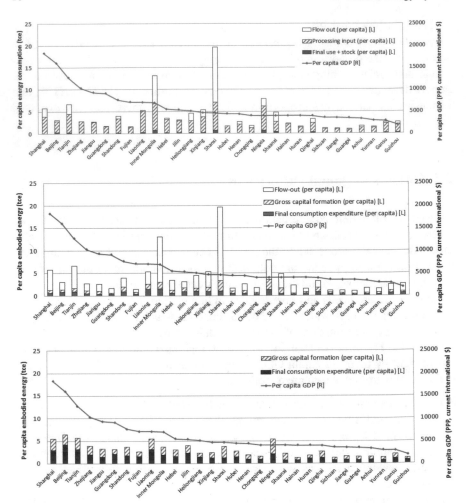

Fig. 2.12 Energy Consumption of 30 Regions by 3 methods (Method 0, A and B from the top to bottom, 2007) (Li et al. 2014) ([*L*] *left* vertical ordinate; [*R*] *right* vertical ordinate)

References

BP (2015) BP Statistical Review of World Energy June 2015

China Electricity Council (CEC) (2008–2013) A brief review table of basic statistics of power industry. http://www.cec.org.cn/guihuayutongji/tongjxinxi/niandushuju/

China Electricity Council (CEC) (2014). The 2014 analyzing and forecasting report on annual situation of electricity supply and demand. http://www.cec.org.cn/guihuayutongji/gongx-ufenxi/dianligongxufenxi/2014-02-25/117272.html

Chong CH, Ma LW, Li Z, Geng J, Zhang TK (2014). A programmed data-processing method for mapping energy allocation Sankey diagram of China. 2014 International Conference on Power and Energy. Taylor & Francis Group, Shanghai, China, pp 28–30

Chong CH, Ma LW, Li Z, Ni WD, Song SZ (2015) Logarithmic mean Divisia index (LMDI) decomposition of coal consumption in China based on the energy allocation diagram of coal flows. Energy 85:366–378

Geng J (2015) Feasibility of natural gas trade between China and Canada based on system modeling and analysis. Thesis of Master of Science on Power Engineering and Engineering Thermophysics, May 2015. Tsinghua University, China

Li Z, Chang SY, Ma LW, Liu P, Zhao LX, Yao Q (2012) The development of low-carbon towns in China: concepts and practices. Energy 47:590–599

Li Z, Pan LY, Fu F, Liu P, Ma LW, Amorelli A (2014) China's regional disparities in energy consumption: an input–output analysis. Energy 78:426–438

Ma LW, Allwood JM, Cullen JM, Li Z (2012a) The use of energy in China: tracing the flow of energy from primary source to demand drivers. Energy 40:174–188

Ma LW, Fu F, Li Z, Liu P (2012b) Oil development in China: current status and future trends. Energy Policy 45:43–53

Ministry of Environmental Protection of People's Republic of China (MEPC). Emission standard of air pollutants from thermal power plants (GB 13223-2011 and GB 13223-2003). http://kjs.mep.gov.cn/hjbhbz/bzwb/dqhjbh/dqgdwrywrwpfbz/index_1.htm

National Bureau of Statistics of China (NBSC) (2012). China Statistical Yearbook 2011. China Statistics Press, Beijing

National Bureau of Statistics of China (NBSC) (2013). 2013 statistical bulletin of national economic and social development. http://www.stats.gov.cn/tjsj/zxfb/201402/t20140224_514970.html

National Bureau of Statistics of China (NBSC) (2014a). China Statistical Year Book 2013. China Statistics Press, Beijing

National Bureau of Statistics of China (NBSC) (2014b). China Energy Statistical Year Book 2013. China Statistics Press, Beijing

National Development and Reform Commission (NDRC) (2014). The Outline of the 10th Five-Year-Plan of Economic and Social Development in China. http://www.sdpc.gov.cn/fzgggz/fzgh/ghwb/gjjh/200709/P020070912638588995806.pdf

National Development and Reform Commission (NDRC) (2014). The Outline of the 11th Five-Year-Plan of Economic and Social Development in China. http://news.xinhuanet.com/ziliao/2006-01/16/content_4057926.htm

National Development and Reform Commission (NDRC) (2014). The 11th Five-Year-Plan of Energy Development. http://www.sdpc.gov.cn/fzgggz/fzgh/ghwb/gjjgh/200709/P020150630514158560149.pdf

National Development and Reform Commission (NDRC) (2014). The Outline of the 12th Five-Year-Plan of Economic and Social Development in China. http://www.sdpc.gov.cn/fzgggz/fzgh/ghwb/gjjh/201109/P020110919592208575015.pdf

NBSC (National Bureau of Statistics of China) and MEPC (Ministry of Environmental Protection of China) (2014). China Statistical Yearbook on Environment 2014. Beijing: China Statistics Press, p 43

People's Net (2014). The database of historical meetings of National Congress of Communist Party of People's Republic of China. http://cpc.people.com.cn/GB/64162/64168/index.html

People's Net (2014). The special plan of energy development in the 10th Five-Year. http://www.people.com.cn/GB/jinji/31/179/20010813/533877.html

Schmidt M (2008) The Sankey diagram in energy and material flow management. J Ind Ecol 12(1):82–94

State Council of China (2014). The 12th Five-Year-Plan of energy development. http://www.gov.cn/zwgk/2013-01/23/content_2318554.htm

Xinhua Net (2014a) Reporting: the annual contribution of coal to PM 2.5 is over 50 %. http://news.xinhuanet.com/energy/2014-10/22/c_127127504.htm

Xinhua Net (2014b) The report of Hu Jintao to the 18th National Congress of Communist Party of China. http://news.xinhuanet.com/18cpcnc/2012-11/17/c_113711665.htm

Chapter 3
Long-Term Forecasting of Macro-Energy Scenarios of China

In the previous chapter we described how China has experienced rapid urbanisation and industrialization growth that has made huge demands on primary energy consumption. The scale and pace of this growth over the past 20 years was not foreseen by economic forecasters in the 1990s. As we look forward to the next 20 years, China's changing economic, demographic, and societal structures introduce further uncertainty in attempting new forecasts. Many renowned energy research institutes periodically publish projections of macro-energy scenarios of China up to 2030 and 2050, but unsurprisingly these projections differ from one another in terms of total amount of energy consumption and energy flows amongst sectors.

In this chapter, we review projections of China's future macro-energy systems up to 2030 and beyond released by international and domestic research institutes. These reports were published between 2010 and 2011 and this enables them to extrapolate China's future energy trends under the same circumstances of the time. It is also important to compare and analyse the different projection methods, key policy assumptions, and other boundary conditions underpinning these different forecasts. Our analysis suggests that projections on energy consumption in China are highly dependent on projections of economic and population growth in most scenarios, whilst in some cases the impacts of oil price, international trade, and other drivers are more influential. It is also notable that projections from domestic research institutes tend to be more optimistic regarding clean and sustainable utilisation of coal in the future.

For our general research we use 2010 as the basis for our long-term forecasts. A major reason for this is availability of data. Selecting 2010 as the base year allows a good balance of coverage of sections and timing of the data. In addition, China's Twelfth Five-year Plan was released in 2011. Whilst setting policy assumptions in our studies, we can apply these policies and examine the effects and implications of them.

© Springer Science+Business Media Singapore 2016
Z. Li et al., *Informing Choices for Meeting China's Energy Challenges*,
DOI 10.1007/978-981-10-2353-8_3

3.1 Introduction

Four major publications form the basis of our analysis

World Energy Outlook 2010 (WEO 2010), published by International Energy Agency (IEA)
International Energy Outlook 2010 (IEO 2010), published by U.S. Energy Information Administration (EIA).
BP Energy Outlook 2030 (BEO 2030), published by British Petroleum (BP).
The Mid-term and Long-term Energy Development Strategy of China (2030, 2050) (MLEDSC), published by the Chinese Academy of Engineering (CAE).

The WEO 2010 describes three scenarios, projecting the world energy trend from 2008 to 2035; the IEO 2010 has set five scenarios, projecting the world energy trend from 2007 to 2035; the BEO 2030 has only one scenario called "most likely scenario", based "to the best of knowledge" rather than "business as usual" extrapolation, projecting the world energy trend to 2030; the MLEDSC uses input from national experts on macro-energy in China, and projecting the energy trend of China from 2010 to 2050. Main features of these reports such as projection horizons, scenarios, and primary considerations are listed in Table 3.1.

Unsurprisingly, these different approaches lead to quite different outcomes, as shown in Figs. 3.1, 3.2, 3.3 and 3.4.

Figures 3.1 and 3.2 illustrate the primary energy demand of China in 2020 and 2035 according to the various scenarios. Obviously, the projections of the primary energy demand in 2020 are closer to each other (Fig. 3.1), while for projections in 2035, we see greater divergence (Fig. 3.2). This is also reflected in the projections of energy-related CO_2 emissions shown in Figs. 3.3 and 3.4.

Table 3.1 Main features of the reports

Reports	Projection horizons	Scenarios	Primary considerations
WEO 2010	2035	Current policies scenario, New policies scenario, 450 scenario	Effects of different policy assumptions on energy consumption
IEO 2010	2035	Reference case, High economic growth case, Low economic growth case, High oil price case, Low oil price case	Effects of different economic growth rates and oil prices assumptions on energy consumption
BEO 2030	2030	Most likely Scenario	Based on "to the best of our knowledge"
MLEDSC	2050	Proposed one scenario	Long-term energy strategies based on potential supply and demand

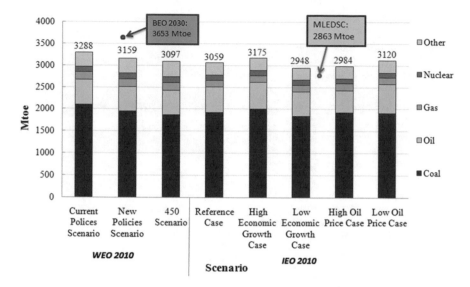

Fig. 3.1 Primary energy demand of China, 2020

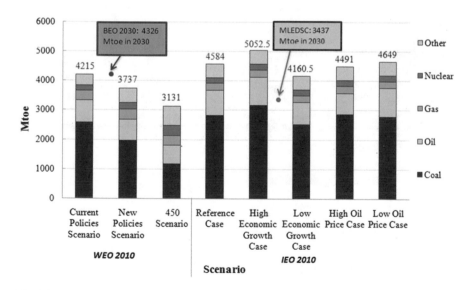

Fig. 3.2 Primary energy demand of China, 2035

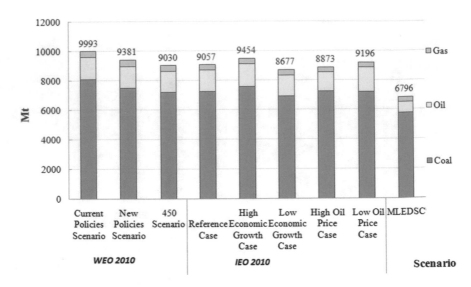

Fig. 3.3 Energy-related CO_2 emission of China, 2020

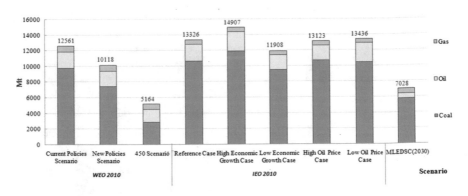

Fig. 3.4 Energy-related CO_2 emission of China, 2035

3.2 Methodology Descriptions

3.2.1 IEA's World Energy Outlook 2010

The IEA has provided medium to long-term energy projections using a World Energy Model (WEM). The model is a large-scale mathematical construct designed to replicate how energy markets function and how the principal tool is used to generate detailed sector-by-sector and region-by-region projections for various scenarios. The model consists of six main modules: final energy demand (with sub-models covering residential, services, agriculture, industry, transport and non-energy use), power generation and heat, refinery/petrochemicals and other transformation, fossil fuel supply, CO_2 emissions and investment.

The model is designed to analyse

Global energy prospects: these include trends in demand, supply availability and constraints, international trade and energy balances by sector and by fuel to 2035.

Environmental impact of energy use: CO_2 emissions from fuel combustion are derived from the detailed projections of energy consumption.

Effects of policy actions and technological changes: alternative scenarios analyse the impact of policy actions and technological developments on energy demand, supply, trade, investments, and emissions.

Investment in the energy sector: the model evaluates investment requirements in the fuel supply chain needed to satisfy projected energy demand to 2035. It also evaluates demand-side investment requirements in the alternative scenarios.

The model structure is shown in Fig. 3.5.

The main exogenous assumptions of the World Energy Model concern economic growth, demographics, international fossil fuel prices, and technological developments. Electricity consumption and electricity prices dynamically link the final energy demand and power generation modules. The refinery model projects throughput and capacity requirements based on global oil demand. Primary demand for fossil fuels serves as input for the supply modules. Complete energy balances are compiled at a regional level and the CO_2 emissions of each region are then calculated using derived carbon factors.

3.2.2 EIA's International Energy Outlook 2010

The IEO2010 projections of world energy consumption and supply were generated from the EIA's World Energy Projections Plus (WEPS+) model. WEPS+

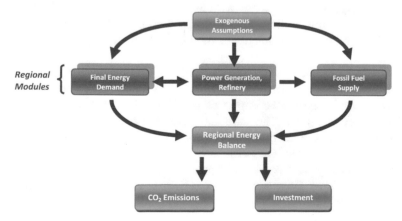

Fig. 3.5 World energy model overview

consists of a system of individual sectoral energy models, using an integrated iterative solution process that allows for convergence of consumption and prices to an equilibrium solution. It is used to build the Reference case energy projections, as well as alternative energy projections based on different assumptions for GDP growth and fossil fuel prices. It can also be used to perform other analyses.

The WEPS+ platform allows the various individual models to communicate with each other through a common, shared database and provides a comprehensive, central series of output reports for analysis. In the individual models, the detail also extends to the subsector level. In WEPS+, the end-use demand models (residential, commercial, industrial, and transportation) project consumption of the key primary energy sources: several petroleum products, other liquids, natural gas, coal, nuclear power, hydropower, wind, geothermal, and other renewable sources. These models also provide intermediate consumption projections for electricity in the end-use demand sectors.

The IEO 2010 projections are based to the extent possible on U.S. and foreign laws, regulations, and standards in effect at the start of 2010. The potential impacts of pending or proposed legislation, regulations, and standards are not reflected in the projections, nor are the impacts of legislation for which the implementing mechanisms have not yet been announced. In addition, mechanisms whose implementation cannot be modelled given current capabilities or whose impacts on the energy sector are unclear are not included in IEO 2010. The projections focus exclusively on marketed energy. Non-marketed energy sources, which continue to play an important role in some developing countries, are not included in the estimates.

3.2.3 BP's Energy Outlook 2030

The BP Energy Outlook is not a "business as usual" extrapolation or an attempt at modelling policy targets. Instead it is built "to the best of our knowledge", reflecting the judgment of the likely path of global energy markets to 2030.

Assumptions on changes in policy, technology, and the economy are based on extensive internal and external consultations. The Policy Case is a fully built-up alternative case, assessing the impact of possible policy changes on energy production and consumption. This case and other sensitivities are used to explore the uncertainties of the Energy Outlook, but do not attempt to forecast long-term energy prices as part of this Outlook.

Historical energy data is fully compatible with the BP Statistical Review of World Energy. Gross Domestic Product (GDP) is expressed in real Purchasing Power Parity (PPP) terms.

3.2.4   CAE's Mid-Term and Long-Term Energy Development Strategy of China (2030, 2050)

The CAE's approach to deriving an overall view of China's energy future was to aggregate the outputs from six sub-projects, including energy saving, coal, gas, nuclear, electricity, and renewable energy. During the project, further research modules relating to the environment, clean coal technology, and hydrogen power research were added to the projection system. More than 40 academics and 200 experts participated in the project. Based on the supply capacity, development potential and energy demand of China, the work summarised and analysed the views and suggestions of these experts, and arrived at an overall view on the targets for future primary fossil energy consumption. Under these targets, the study took into account constraints associated with China's natural resources and the environment. The study provided an important perspective on China's possible development speed, industrial structure, power mix, and consumption, and therefore a view on a sustainable energy development pathway for China. As a result, the CAE study looked towards policy shaping a future that was both achievable and desirable in contrast to the IEA, EIA, and BP studies, which look at what the likely outcome could be and then test different policy scenarios, primarily for carbon mitigation.

3.3   Policy Assumptions

Policies play significant role in scenario analysis, and have great impact on energy demand side and supply side, as well as the energy market. The four institutions set policies to test different scenarios and compare the different outcomes. In WEO 2010, the policies settings in the three scenarios are mainly about carbon mitigation. Thus, policies finally influence the carbon emission in each scenario. In IEO 2010, the policies are based to the extent possible on U.S. and foreign laws, regulations, and standards in effect at the start of 2010, while the scenario analysis is mainly determined by economic growth rate and oil prices. In BEO 2030 and MLEDSC, the policy changes assumptions are based on internal and external consultations. The details of policy assumptions are listed in Table 3.2.

3.4   Technical Accuracy

The four references differ not only in primary energy demand but also categories used for energy resources and consumption sectors, as shown in Figs. 3.6 and 3.7.

Figure 3.6 illustrates the primary energy categories in the scenarios of the four institutions. Coal, gas, and nuclear are separately categorised. The three

Table 3.2 Policy assumptions of the scenarios

Report	Scenario	Policies
WEO 2010 (IEA)	Current policies scenario	Serve as a baseline against which the impact of new policies can be assessed No change in policies is assumed: – Takes into account those measures that governments had formally adopted by the middle of 2010 in response to and in pursuit of energy and environmental policies – Takes no account of any future changes in government policies – Does not include measures to meet any energy or climate policy targets or commitments that have not yet been adopted or fully implemented
	New policies scenario	**Overall targets and policies:** – 40 % reduction in CO_2 intensity by 2020 compared with 2005 (2009) – Rebalancing of the economy from industry towards services (2009) – Further implementation of the directives of the Renewable Energy Law (2005) **Detailed sectoral policies of power, transport, industry and building sectors**
	450 scenario	**Overall targets and policies:** – 45 % reduction in CO_2 intensity by 2020 compared with 2005 – 15 % share of non-fossil energy in primary energy consumption by 2020 **Detailed sectoral policies of power, transport, industry and building sectors**
IEO 2010 (EIA)		– Based to the extent possible on U.S. and foreign laws, regulations, and standards in effect at the start of 2010 – The potential impacts of pending or proposed legislation, regulations, and standards are not reflected in the projections, nor are the impacts of legislation for which the implementing mechanisms have not yet been announced – Mechanisms whose implementation cannot be modelled given current capabilities or whose impacts on the energy sector are unclear are not included – IEO2010 focuses exclusively on marketed energy – Non-marketed energy sources, which continue to play an important role in some developing countries, are not included in the estimates

(continued)

Table 3.2 (continued)

Report	Scenario	Policies
BP energy outlook 2030 (BP)	"To the best of our knowledge"	Assumptions on changes in policy, technology and the economy are based on extensive internal and external consultations
The mid-term and long-term energy development strategy of China ("MLEDSC" for short) (CAE)		– Save energy and control the total energy consumption – Utilise coal in a scientific, clean and high-efficient way – Assure the strategic positions of oil and natural gas, consider natural gas as one of the key resources to adjust the energy structure – Accelerate development of hydro power and other renewable energy – Take great efforts to develop nuclear power – Develop smart-grid systems

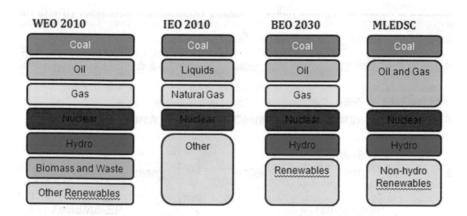

Fig. 3.6 Primary energy categories

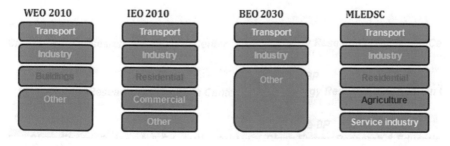

Fig. 3.7 Energy consumption categories

non-Chinese institutes consider oil (or liquids) as a separated energy resource, while CAE combines oil and gas together to analyse the demand. Hydro power is separately analysed except in the IEO 2010 scenarios. Only WEO 2010 takes account of biomass and waste as a part of the renewable energy demand.

Figure 3.7 illustrates the energy consumption categories in the scenarios of the four institutions. Each institution has slight differences in analysing the energy consumption sectors. Besides transport and industry, WEO 2010 considers energy consumption for buildings, IEO 2010 and MLEDSC consider residential energy consumption, IEO 2010 considers commercial energy consumption and MLEDSC considers energy consumptions of agriculture and service industry.

3.5 Criteria of Scenario Settings

The four reports discussed here have complicated assumptions and settings for their scenario designs, including population, economic growth, oil and carbon price, technology development, as well as policy assumptions. The key criteria that influence the scenario analysis are listed in Table 3.3.

As shown in Fig. 3.8, the WEO 2010, IEO 2010 and MLEDSC have set specific population growing trend of China for the scenario analysis. In WEO 2010, China population is assumed to grow at an annual average rate of 0.6 % from 2008 to 2020, and 0.1 % from 2020 to 2035. The annual average growth rate from 2008 to 2035 is 0.3 %. The rates of population growth assumed in all three scenarios are based on the most recent projections by the United Nations (UNPD, 2009). Population growth slows progressively, in line with past trends. In IEO 2010, China's population is projected to be 1421 million in 2020, 1452 million in 2035, and the annual growth rate is 0.3 % from 2007 to 2035. In MLEDSC, the annual population growth rate of China is 4.5 % from 2010 to 2030, nearly flat from 2030 to 2040, and −2.5 % from 2040 to 2050. The annual average growth rate is 0.36 % from 2010 to 2035, and 0.16 % from 2010 to 2050.

Figure 3.9 illustrates the GDP growing trend of China in varied scenarios. In MLEDSC, the GDP of China grows much faster than that in WEO 2010 and IEO 2010. In WEO 2010, China GDP is assumed to grow at an annual average rate of 7.9 % from 2008 to 2020, and 3.9 % from 2020 to 2035. The annual average growth rate is 5.7 % from 2008 to 2035. In IEO 2010, the annual average growth rates from 2007 to 2035 range from 5.7 to 6.2 depending on different scenarios, as listed in Table 3.2. In MLEDSC, the annual growth rate is assumed to be 8 % from 2010 to 2030, and higher than 4 % from 2030 to 2050.

Besides the criteria settings, there are detailed descriptions on scenario settings for each of the scenario, as listed in Table 3.4.

The WEO 2010 and IEO 2010 especially focus on the impacts of oil prices on the energy market, and the oil price assumptions are shown in Fig. 3.10. The oil prices assumed in IEO 2010 are higher than that in WEO 2010 scenarios.

Table 3.3 Criteria of scenario settings of China

Criteria	Year	WEO 2010 (IEA)	IEO 2010 (EIA)					BP Energy outlook 2030 (BP)	MLEDSC (CAE)
			Reference case	High economic growth scenario	Low economic growth scenario	High oil price scenario	Low oil price scenario		
Length of prediction period		2008–2035	2007–2035					2010–2030	2010–2050
Population (millions)	2020	1421	1415						1394
	2030	1442	1429						1458
	2035	1452	1437						1458
	2050								1422
Annual growth rate (%)	2010–2035	0.34	0.30						0.36
GDP (billion 2005 dollars)	2020	17,969	17,353	18,264	16,483	17,204	17,499		18,136
	2030	26,344	24,709	27,418	22,362	24,627	24,936		39,155
	2035	31,898	32,755	37,039	28,950	32,493	33,056		47,638
	2050								85,793
Annual growth rate (%)	2010–2035	5.7	5.8	6.2	5.3	5.7	5.8		7.2

Fig. 3.8 Population growth setting (China)

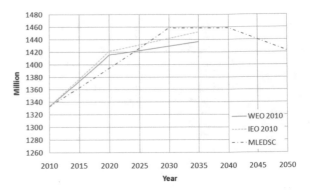

Fig. 3.9 GDP growth setting (China)

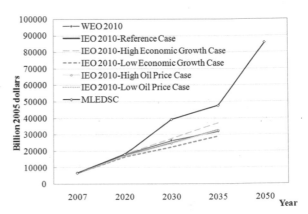

The WEO 2010 also set carbon prices in its three scenarios, acting as emission constrains for fossil energy consumptions. The CO_2 prices by main region and scenario are listed in Table 3.5.

3.6 Challenges of Projections and Our Motivations

In the previous sections, we have highlighted the amount of detailed analysis that goes into the forecasting of future energy demand. We have also shown how different economic, technological, societal and policy assumptions, and modelling approaches, will unsurprisingly lead to variances between different studies.

The absolute figures are not critical. China will need more energy as its economy grows. All the studies suggest that from 2020 to 2035 that China will consume around 50 % more primary energy at the end of that period. We cannot be certain of this but history has shown us that energy forecasting can underestimate the future growth.

Table 3.4 Scenario settings

Report	Scenario	Descriptions
WEO 2010 (IEA)	Current policies scenario	• Takes into account those measures that governments had formally adopted by the middle of 2010 in response to and in pursuit of energy and environmental policies • Takes no account of any future changes in government policies • Does not include measures to meet any energy or climate policy targets or commitments that have not yet been adopted or fully implemented
	New policies scenario	Takes account of the broad policy commitments and plans that have been announced, yet does not assume that the policy commitments are all fully implemented
	450 scenario	• The concentration of greenhouse gases in the atmosphere is limited to around 450 parts per million of carbon dioxide equivalent • Reflects an assumption of vigorous policy action to implement fully the Copenhagen Accord
IEO 2010 (EIA)	Reference case	A business-as-usual trend estimate, does not include prospective legislation or policies
	High economic growth case	Consider the effects of higher and lower growth paths for economic activity than are assumed in the Reference case
	Low economic growth case	
	High oil price case	The impacts of world oil prices on energy demand are a considerable source of uncertainty in the IEO 2010 projections. In addition to the Reference case, High Oil Price and Low Oil Price cases illustrate the range of that uncertainty, although they do not span the complete range of possible price paths
	Low oil price case	
BP energy outlook 2030 (BP)	"To the best of our knowledge"	Reflect the judgment of the likely path of global energy markets to 2030

(continued)

Table 3.4 (continued)

Report	Scenario	Descriptions
The mid-term and long-term energy development strategy of China (CAE)		Based on the situation and key constrains of the mid-term and long-term energy development of China, systematically research on the supply capacity, development potential and reasonable demand of the key energy resources, and provide the strategy thoughts, targets, focal points, pathways, technological supports, and policy suggestions of the energy development

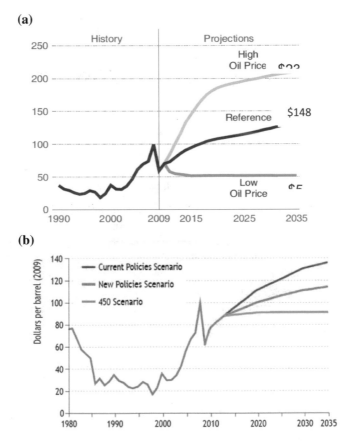

Fig. 3.10 Oil price assumptions (2009 dollars per barrel). **a** IEO 2010 assumptions on oil price. **b** WEO 2010 assumptions on oil price

Table 3.5 CO_2 prices by main region and scenario ($ per tonne)

Region		2009	2020	2030	2035
New policies	European Union	22	38	46	50
	Japan	n.a.	20	40	50
	Other OECD	n.a.	–	40	50
Current policies	European Union	22	30	37	42
450	OECD+	n.a.	45	105	120
	Other major economies	n.a.	–	63	90

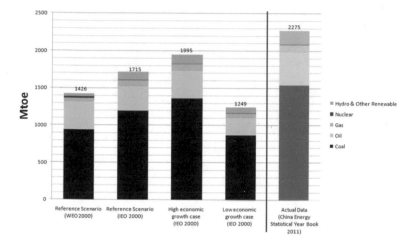

Fig. 3.11 Comparisons of projections and actual data of China's primary energy consumption in 2010

If we look at projections of China's energy consumption made back in 2000 and compare to actual statistics, the differences are significant as shown in Fig. 3.11. For instance, WEO 2000s projection fell way short, only predicting around two thirds of the actual consumption. Although a high economic growth case was considered in IEO 2000, the projection was still lower than the actual. It is this huge potential for growth that the rest of this book will begin to focus on. Whether it is 50, 60 or 40 % greater in 2035 relative to 2020, China needs to find an efficient and sustainable path to satisfy this future energy demand.

In our research, rather than focusing on extrapolating future trends for China's energy demand and consumption, we try to give insights into factors that may alter future trends for the energy system, and suggest approaches for optimising future energy pathways, particularly in the power sector.

Chapter 4
China's Fossil Fuel Resources

Fossil fuels dominate primary energy globally and this is no different in China. In this chapter, we provide an overview of China's fossil fuel resources: the reserves, production, consumption, and the imports of coal, gas, and oil. Unconventional gas (for example coal-bed methane (CBM) and shale gas) is an important potential resource to meet China's huge appetite for energy and if gas can begin to offset coal use then it can help in China's battle to reduce its carbon emissions while still growing economically. However, we identify several barriers that slow rapid exploitation of this unconventional resource and suggest some policies to break down these barriers.

4.1 Overview

4.1.1 Coal

4.1.1.1 Coal Reserves

According to the latest evaluation of national coal resources conducted by the Ministry of Land and Resources in 2014 (MLR 2014), the total prospective resources of China's coal at a depth shallower than 2000 m were about 3.88 trillion metric tonnes and the discovered reserves of coal were 1.48 trillion metric tonnes, accounting for 38.1 % of the prospective resources by the end of 2013. Most coal reserves are located in the north and north-west of China, which poses a large logistical problem for supplying electricity to the more heavily populated coastal areas (Fig. 4.1).

© Springer Science+Business Media Singapore 2016
Z. Li et al., *Informing Choices for Meeting China's Energy Challenges*,
DOI 10.1007/978-981-10-2353-8_4

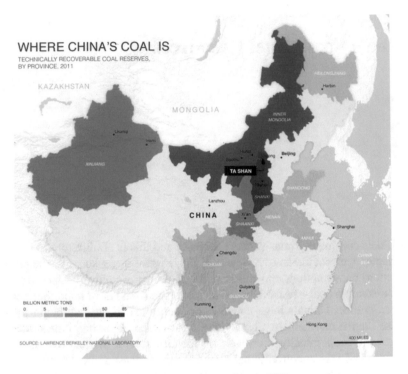

Fig. 4.1 Technically recoverable coal reserves by province in 2011

4.1.1.2 Coal Balance

Coal is the dominant primary energy in China and China produces the largest amount of coal in the world.

China's annual coal production increased from 1080 million metric tonnes as of 1990 to 3974 million metric tonnes as of 2013, with an average annual growth rate of 5.7 % (NBS 2013). China has become a net importer of coal after 2008. But the net import of coal is small; in 2013 it was 319 million metric tonnes, only accounting for 7.5 % of coal consumption in the same year.

China consumed 4244 million metric tonnes of raw coal in 2013, accounting for 67.4 % of the total primary energy consumption. A sizeable proportion 1898 million metric tonnes of coal were used for power generation, accounting for 45 % of the total national coal consumption (Table 4.1).

The consumption of coal is largely in power production, aside from this, there is a significant amount used in industry and manufacturing but a comparatively small amount of domestic use. China's installed power capacity of pulverised coal (PC) was 862 GW, or 69 % of the total power capacity, in 2013. China's energy consumption is mostly driven by the industry sector, the majority of which comes from coal consumption. One of the principal users is the steel industry in China. In the largest cities, such as Beijing, the domestic burning of coal is no longer

Table 4.1 Coal balance of China (1990–2013) (unit: million metric tonnes)

Item	1990	1995	2000	2005	2006	2007	2008	2009	2010	2011	2012	2013
Productions	1080	1361	1384	2365	2570	2760	2903	3115	3428	3764	3945	3974
Imports	2	2	2	26	38	52	44	132	183	222	288	327
Exports (−)	17	29	55	72	63	53	46	22	19	15	9	8
Stock changes	−42	1	−12	35	70	66	46	−9	−37	−42	−38	−44
Consumption	1055	1377	1357	2434	2706	2904	3006	3250	3490	3890	4117	4244

Data source National Bureau of Statics. China Energy Statistical Yearbook 2014

permitted. In rural areas coal is still permitted to be used by Chinese households, commonly burned raw in unvented stoves.

Sidebar: Definitions of Resources and Reserves in China

1. Prospective resources: the potential amount, that is, the probable maximum that can be exploited. In the assessment of resources conducted by the Ministry of Land and Natural Resources, prospective resources have a probability of 5 %.
2. Geological resources: the amount that can be discovered using current techniques, including proven and unproven resources.
3. Recoverable resources: the amount of resources that can be exploited with future foreseeable technologies and economic conditions.
4. Proven reserves: under the specified economic, technological, and policy conditions, the developed and undeveloped reserves.
5. Remaining proven resources: for developed oil or gas reserves, the remaining proven resources are equal to the recoverable resources minus the amount of extracted oil or gas.

4.1.2 Gas

4.1.2.1 Gas Reserves

Based on the latest evaluation of conventional oil and gas resources conducted by the Ministry of Land and Resources in 2014 (MLR 2014), the geological resources of China's conventional gas were about 62 trillion m^3, the recoverable resources were nearly 40 trillion m^3, and the remaining recoverable resources were 4.64 trillion m^3 by the end of 2013.

The proven reserves of conventional gas have continuously risen in China, as shown in Fig. 4.2 (BP 2014). However, accumulated proven reserves were 11.4 trillion m^3 by the end of 2013, only accounting for 18.4 % of the geological resources. Therefore, China's gas is at the early stage of exploration, and has great potential for further exploration and development.

Moreover, China has abundant CBM resources: the reserves of CBM at a depth shallower than 2000 m were about 36.8 trillion m^3, the recoverable resources were about 10.8 trillion m^3, and proven reserves were 273.4 billion m^3 by the end of 2013 (MLR 2014).

4.1.2.2 Gas Balance

In 2013, the production of gas in China was 121 billion m^3, rising 9.8 % over levels of the year before, when the production of conventional gas was 117.8 billion m^3, the production of shale gas was 0.2 billion m^3, and the production of CBM was

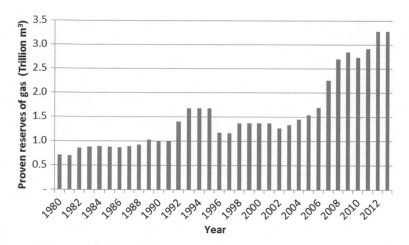

Fig. 4.2 Proven reserves of conventional gas in China (1980–2013) (BP 2014)

3 billion m³. The import of gas was 53.4 billion m³, increasing 25.6 % over levels of the previous year. The apparent consumption of gas was 169.2 billion m³, rising 12.9 % over the levels of the prior year, 2012.

China's annual natural gas production increased from 15 billion m³ as of 1990 to 121 billion m³ as of 2013, with an average annual growth rate of 9.3 % (NBS 2013).

The import of natural gas has increased in China in recent years. The net gas import of 2013 was 50 billion m³, accounting for 29.2 % of gas consumption in the same year. Importation of natural gas is mainly completed by LNG terminals and onshore pipeline. In 2009, China imported about 7.63 billion m³ of LNG (BP 2010), equivalent to 5.6 million metric tonnes of LNG, mainly from the already completed LNG import projects in Guangdong and Fujian. Onshore import pipeline projects, including the Central Asia—China Gas Pipeline, Russia—China pipeline, and Myanmar—China pipeline, will be completed over time. However, China's imports of natural gas are restricted by the high costs related to transportation of gas over significant distances.

The demand for natural gas grew rapidly between 2000 and 2013, and the natural gas consumption in China expanded 6.7 times; in 2013, China's natural gas consumption reached 171 billion m³. The average annual growth of natural gas consumption reached 16.2 % and it is far higher than the growth rate of energy consumption (9.4 %) over the same period (Table 4.2).

China's natural gas is mainly consumed for industrial purposes, like chemical gas and industrial fuel, only a small proportion is used as city gas or for power generation, and a smaller proportion is used as automobile fuel (see Table 4.3). In recent years, demand for natural gas in each sector has grown rapidly. The structure of natural gas consumption has changed: the proportion of chemical gas and industrial gas has begun to decline, the proportion of city gas and power generation has increased rapidly, and transportation gas increased to about 4 %.

Table 4.2 Gas balance of China (1990–2013) (unit: billion cubic metres)

Item	1990	1995	2000	2005	2006	2007	2008	2009	2010	2011	2012	2013
Production	15	18	27	49	58	69	80	85	96	105	111	121
Imports					1	4	4	7	16	31	42	53
Exports (−)			3	3	3	3	3	3	4	3	3	3
Consumption	15	18	24	46	56	70	81	89	108	134	150	171

Data source National Bureau of Statics. China Energy Statistical Yearbook 2014

Table 4.3 Structure of natural gas consumption in China

Years	2000		2005		2008		2013	
	0.1 billion m^3	%	0.1 billion m^3	%	0.1 billion m^3	%	0.1 billion m^3	%
Chemical industry	82.3	33.6	141.4	30.2	200.0	24.6	305	17.9
Power generation	6.4	2.6	18.8	4.0	73.9	9.1	244	14.3
Industrial fuel	110.3	45.0	168.6	36.1	257.7	31.7	580	34.0
Transportation[①]	5.8	2.4	19.7	4.2	32.7	4.0	176	10.3
City gas[②]	40.2	16.4	119.1	25.5	248.6	30.6	400	23.5
Total	245.0	100.0	467.6	100.0	812.9	100.0	1705	100

Note [①]Gas consumption in transportation is sector statistics
[②]The sum of civil, commercial and other consumption
Data source National Bureau of Statics. China Energy Statistical Yearbook 2014

4.1.3 Oil

4.1.3.1 Oil Reserves

Based on the latest evaluation of conventional oil and gas resources conducted by the Ministry of Land and Resources in 2014 (MLR 2014), the geological resources of China's conventional oil were about 103.7 billion metric tonnes, the recoverable resources were nearly 26.8 billion metric tonnes, and the remaining recoverable resources were 3.37 billion metric tonnes by the end of 2013.

The proven reserves of conventional oil remain relatively flat in China over the past 30 years, as shown in Fig. 4.3 (BP 2014). The accumulated proven reserves were 34 billion metric tonnes by the end of 2013, accounting for 33 % of the geological resources. Although the conventional oil reserve-production ratio was as low as 12:1 in 2013 (BP 2014), China still has considerable potential for expanding its oil reserves.

In addition, recoverable resources of unconventional oil resources, such as oil shales and oil sands, are estimated to be 12 and 2.3 billion metric tonnes, respectively; these can be important supplements to conventional resources.

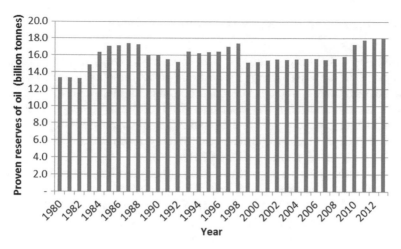

Fig. 4.3 Proven reserves of conventional oil in China (1980–2013) (BP 2014)

4.1.3.2 Oil Balance

From 1995 to 2013, China's crude oil production showed a slight increase of only 60 Mt; by contrast, its oil imports increased by 264 Mt, resulting in its oil import dependency (OID) increasing dramatically from 6.6 to 58 % (Table 4.4). China became a net oil-importing country after 1995, and the amount it imports has been increasing ever since. In 2013, China surpassed the US to become the world's largest importer of oil.

The strategic petroleum reserves (SPR) is one of the basic measures for ensuring energy security for oil-importing nations. In 2004, China proposed a three-stage plan to establish its national SPR. It targets building a reserve of 85 Mt, which can sustain the country's oil consumption for 100 days, which is slightly higher than the standard set by the IEA. The whole plan is expected to be completed by 2020. The statistical data shows that the national SPR in 2010 exceeded 23 Mt (NBS 2011).

Table 4.4 Crude oil balance of China (1990–2013) (unit: million metric tonnes)

Item	1990	1995	2000	2005	2006	2007	2008	2009	2010	2011	2012	2013
Production	138	150	163	181	185	186	190	189	203	203	207	210
Imports	3	17	70	127	145	163	179	204	238	254	271	282
Exports (−)	24	18	10	8	6	4	4	5	3	3	2	2
Stock changes	1	−1	−9	1	−1	−5	−10	−7	−9	−14	−9	−3
Consumption	118	149	212	301	323	340	355	381	429	440	467	487

Data source National Bureau of Statics. China Energy Statistical Yearbook 2014

4.1.4 Future Fossil Energy Development Trend

Overall, Fossil fuel energy satisfies most of China's energy demand, accounting for about 90 % of China's energy production and consumption (see Fig. 4.4). Coal is the dominant energy source in China, accounting for more than 75 % of China's primary energy production and more than 65 % of primary energy consumption. Oil and natural gas accounted for 17.1 and 5.3 % of China's primary energy consumption in 2013, nuclear and renewable energy accounted for only 11.8 % of China's primary energy consumption in 2013.

In the future, China's energy structure must become cleaner and more environmentally friendly. Therefore, fossil fuel will be replaced by clean energy (nuclear, renewable or other energy) gradually. However, the fossil fuel will remain the dominant primary energy of China for a long time, because other types of energy cannot meet the huge energy demand of China.

Natural gas is a clean, efficient, and relatively environmentally friendly fossil fuel, and the specific CO_2 emission of gas-fired power is only half of that of coal-fired power. So fossil fuel can become cleaner by switching coal to gas. Moreover, gas only accounts for a very small part of primary energy consumption in China (5.3 % in 2013), comparing the worldwide average level (23.7 % in 2013). Therefore, gas can be expected to play a major role in China's energy consumption and the demand of gas in China should become stronger.

However, studies show that China's peak annual output of conventional gas could only reach 240–280 billion m^3 (Lu 2009), which is insufficient to meet domestic gas demand. This mismatch between conventional gas supply and demand is therefore a chronic problem. However, China has huge reserves of unconventional gas (CBM, shale gas, tight gas). CBM has been commercially produced for several years and shale gas production is being developed in China today. Therefore, development of this unconventional gas resource is an important option for a cleaner energy future.

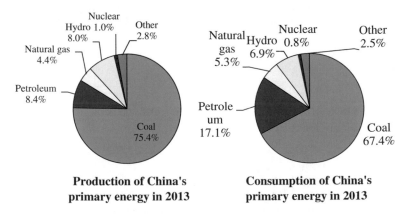

Fig. 4.4 Production and consumption of China's primary energy in 2013. *Source* National Bureau of Statics. China Energy Statistical Yearbook 2014

4.2 Unconventional Gas—Challenges for This Untapped Resource

In recent years, the exploration and development of natural gas, especially unconventional gas, has received great attention in China. Shale gas has recently been listed as an independent mineral resource, and a series of preferential policies on the exploration and development of CBM and shale gas have been promulgated; besides, an initial public bidding of the exploration and development of shale gas has been conducted in a well-organised manner.

However, the conflict of mineral rights, which always slowed the expansion of CBM, has yet to be resolved, and the true economic viability and environmental constraints of shale gas extraction are, as yet, untested at scale.

At the same time, China's oil and gas market structure creates issues that slow development and fails to attract new investors and investment. The pricing mechanism in China's natural gas market, the power of the traditional state-owned enterprises (SOEs), and limited transmission infrastructure and existing pipeline access all hinder the development of a booming unconventional gas industry. Some of these challenges are not unique to China. Only the United States in the last decade has been successful at exploiting shale gas resources, and it has done so because of an open approach to a dynamic entrepreneurial energy industry.

Acting to overcome these domestic impediments to the development of unconventional gas is vital to the development of unconventional gas industry in China.

4.2.1 Definition and Classification

Unconventional gas resources mainly include tight sandstone gas, CBM, and shale gas. Tight sandstone gas relates to a low permeability gas which is stored in sandstone. CBM is a self-generation and self-storage natural gas, which formed and remained in the coal seam. Shale gas is a natural gas that is stored in shale rock. These unconventionals share common characteristic, low reservoir permeability. Because of the low permeability, the gas production is very low using conventional development techniques. Another definition of unconventional gas relates to gas reservoirs whose production cannot bring economic benefit using common technologies unless stimulation methods (such as fracturing, horizontal wells or cluster wells) are adopted.

Figure 4.5 is the triangle diagram of unconventional gas resources. The scale on the right side of Fig. 4.5 shows the standard value of gas reservoir permeability. The scale on the left side indicates the average development cost of different gas resources. The volume means the potential gas reservoirs. From the diagram, it can be seen that there is considerable economic production potential for unconventional gas resources for the foreseeable future. The reservoir quality, i.e. the reservoir permeability, decreases progressively from top to bottom but their quantity

Fig. 4.5 Triangle diagram of gas resources. *Source* Oil & Gas Journal

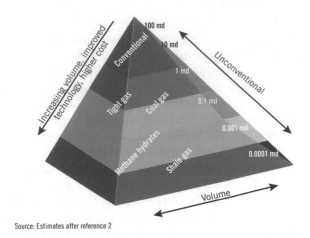

Source: Estimates after reference 2

increases in the diagram. Different resources of unconventional gas have different development costs. The development cost of shale gas is usually higher than that of tight gas and CBM, while the development of methane hydrates is the most expensive and still in its infancy (see Fig. 4.5).

4.2.2 Development Status

4.2.2.1 CBM

China is rich in CBM resources. According to the latest assessment of CBM resource in China, CBM geological resource with a depth within 2000 m is about 36.8×10^{12} m^3, equal to onshore conventional natural gas resources, ranking third in the world. Geological resource density is 0.98×10^8 m^3/km^2 and they are mainly distributed in Northern China and the Northwest. As of the end of 2010, there are 2139.53 km^2 of proven area of CBM, 2863.36×10^8 m^3 of proven geological reserves, and 1480.9×10^8 m^3 of proven recoverable reserves.

As of the end of 2009, 113 effective CBM rights (exploration right or mining right) were granted by the Ministry of Land and Resources of China and the total area was 65,520 km^2. There were 106 CBM exploration rights with a total area of 65,285 km^2, distributing in 17 provinces. There were 7 CBM mining rights with a total area of 235 km^2, distributed in Shanxi, Liaoning and Shaanxi provinces.

By the end of 2009, an underground CBM exhaust system was established in nearly 200 coal mines in China, realising an exhaust volume of 64.5×10^8 m^3. The change of underground exhaust volume of CMM (coal mine methane) from 2000 to 2009 is shown in Table 4.5. There are ten mining groups (including

Table 4.5   CBM production in China (unit: 108 m^3)

Year	2000	2001	2002	2003	2004	2005	2006	2007	2008	2009
Underground exhaust	9	10	12	15	19	23	32	44	48	64.5
Surface exhaust	–	–	–	–	–	03	1.3	3.8	7.5	10.1

Table 4.6   CBM production in China (unit: 10^8 m^3)

Year	Duanshi-Qinshui	Duanshi-Jincheng-Bo'ai	Jincheng
Transportation capacity (10^8 m^3)	30	10–20	10
Length (km)	35	98	45

Yangquan, Huainan, Shuicheng, Panjiang, Songzao, Jincheng, Fushun, Huaibei) having an annual extraction volume over 1×10^8 m^3.

By the end of 2009, 180 km of CBM transmission pipeline had been established (see Table 4.6). Eight small-scale CBM CNG stations had been built, having a combined compression capacity of 130×10^4 m^3. Three CBM LNG stations were built and the liquefaction capacity is 155×10^4 m^3.

4.2.2.2   Shale Gas

In 2012 Shale gas became China's 172nd defined mineral product. A preliminary estimation of shale gas resources has been made for major basins and regions in China. The results indicate that the shale gas resources in major basins and regions of China are circa 15–30×10^{12} m^3, with a median value of 23.5×10^{12} m^3, roughly equivalent to the 28.3×10^{12} m^3 in USA. The Sichuan Basin and Tarim Basin are large sedimentary basins, abounding in organic matter-rich shale, thus they have good prospects for shale gas exploitation. In addition, there are an additional five large-scale shale gas basins in China, but their recovery prospects are not clear (see Fig. 4.6).

4.2.3   Development Barriers

For China to exploit these resources, several regulatory and commercial barriers need to be overcome. China's shale resources are believed to be more complex than those in the United States and also unlike the United States distant from the consuming markets. Therefore much more government effort is required to stimulate the development of this resource.

Lessons from the United States suggest that ready access to transmission infrastructure and the market is one of the key elements of success. In China, such access is limited, partly due to geography and partly due to the domestic market

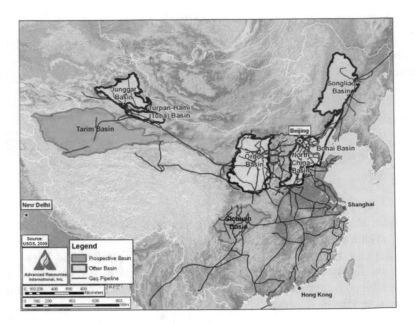

Fig. 4.6 Major shale gas basins and pipelines in China

structure. Since the large Chinese state-owned enterprises (SOEs) control the pipeline network, this leads to a risk of monopolistic control and reduced access to
competitors. There is also a risk of duplication of infrastructure and investment by
these SOE's, as they pursue their own business growth strategies.

The cost-plus pricing mechanism for natural gas results in: (1) Price of natural gas cannot reflect the market supply and demand, (2) an uncompetitive price
between natural gas and other fuels and (3) a significant price imbalance between
imported natural gas and domestic natural gas. As a result, the project risk-reward
profile is unattractive to new entrants.

The regulatory system for a nationwide, expanded natural gas industry is
immature, with a lack of specialised regulatory agencies and legal framework for
regulation. In addition, the number of government departments involved makes it
difficult to achieve a swift and unified regulatory system.

Some of the problems being faced by the fledgling Shale Gas industry have
been experienced by the older CBM sector. China has struggled to overcome these
issues.

Lack of investment and financing continues to be an ongoing issue for CBM
with Banks reluctant to lend. Foreign investment has not made a major breakthrough, with foreign governmental and environmental protection funds not available for coal related developments because of the environmental concerns.

The natural geological overlap of CBM and coal creates additional complexity
over exploitation rights. The mineral rights of CBM are controlled by the national
Ministry of Land and Resources, while most of the exploration and mining rights

of coal are controlled by the local provincial governments. This can create misalignment of interests.

CBM exploitation and utilisation have therefore remained limited. CBM gas quality also makes it difficult to be transported and utilised across a gas network. The limited growth of gas power and low market price of gas come together to depress large-scale developments of CBM.

Hence it is no surprise that at this early stage of the shale gas industry, the development costs for shale gas are high and the rate of economic return on investment is too low. Therefore it is difficult to encourage significant investment from companies with the relevant know how. It is critical that China attracts the right operating companies to deliver and manage these resources and avoid environmental issues, such as groundwater contamination by chemicals in hydraulic fracturing liquids.

4.2.4 Recommendations

In our review of the unconventional gas industry in China, we have concluded there is a need to stimulate the development of China's natural gas resource. The lessons from CBM in China and the success of the United States in the exploitation of Shale have provided a rich list of recommendations. These are listed below.

4.2.4.1 Natural Gas Industry

Regulations for natural gas pipeline

1. Establish a national natural gas pipeline transmission company. The provincial pipeline networks should be absorbed into this entity to create a unified and efficient operation.
2. Establish fair, open, independent oversight, and supervision mechanism for this entity. The main priorities are to ensure a fair third party access policy, control of the pipeline transportation rate and construction and supervision of the system.

Regulations for natural gas pricing mechanism

1. Introduce rational and fair natural gas pricing. Ahead of any free market pricing, the method should be based on netback pricing, allowing reasonable profit, and consistent with calorific value price of alternative fuel (market substitution) for reference.
2. Improve the price supervision and clearing system. China should make and set up a natural gas price regulation system, providing oversight and establish basic methods that are clear and legally binding.

3. Separate the functions between the policy making and policy implementation
4. Set up quasi-governmental regulatory agencies, which belong to the government departments but are relatively independent of the government.
5. Unify regulatory principles. Regulators should follow the principle of openness, the principle of transparency, consistency, responsibility, and independence.

4.2.4.2 CBM Sector

CBM investment and financing channels

1. Clarify the strategic role of CBM and increase the national investment.
2. Create preferential policies on taxes, subsidies, and environmental protection to attract capital investment.
3. Open up to international financing channels. The foreign environment fund utilisation rate must be improved without affecting the diplomatic principles.

Solution for overlapping CBM and coal mining rights
The resource allocation principle of CBM mining priority should be given

1. Promote a "CBM priority", especially for large-scale undeveloped coal fields (also CBM fields). The allocation time of CBM mining rights should be 10–20 years earlier than the coal mining rights.
2. Promote a "Complete set principle". For large-scale integrated coal fields, the CBM mining rights should be sold completely, instead of a number of small configurations.
3. Promote a "Commercializing configuration". Open tender or auction should be configured by the market, rather than through administrative transfer.

Expanding the utilisation of CBM

1. Revise the CBM power policy, and strengthen the supervision and management of power grid enterprises.
2. Revise "coal mine safety regulations", and develop utilisation technologies for low concentration CBM.
3. Cultivate a CBM consumer market. A flexible market investment mechanism and CBM distribution system should be established.

4.2.4.3 Shale Gas Sector

Establishing preferential policies for shale gas finance and taxation

1. Based on the successful expansion of shale gas in countries like the United States, introduce similar preferential policies for shale gas as for domestic CBM exploration and development.

Strengthening technology innovation and basic exploration work

1. China should introduce, absorb, innovate, and improve the shale gas reservoir evaluation techniques, perforation optimisation technology, horizontal well technology and fracturing technique, and gradually create a number of suitable indigenous technologies, suitable for Chinese shale gas geological characteristics.
2. For the exploration work, a plan for shale gas resources investigation, exploration, and development should be made. Geological theory and new understanding for local shale gas should be studied, and a comprehensive nationwide shale gas geological survey carried out.

Establishing regulations of environmental protection

1. Apply strict environmental protection laws and regulations to force the shale gas production enterprises to take corresponding environmental protection measures.

4.3 Conclusions

Fossil fuel energy satisfies most of China's energy demand, accounting for more than 90 % of China's energy production and consumption. Coal is the dominant primary energy in China, accounting for about 70 % of China's energy production and consumption. The proven reserves of coal can support the energy demand of China for 31 years, but the exploration and utilisation of coal should also be mindful of its impact on local pollution and CO_2 emissions.

However energy security continues to be a huge concern for the country. China is not rich in conventional oil and gas resources. The OID reached 58 % in 2013 to meet the huge oil demand in China. Unconventional natural gas has been less explored in China than other energy sources and offers the chance of cleaner burning fuel from domestic, reserves of coal-bed methane and shale gas.

References

BP (2010) Statistical review of world energy 2010
BP (2014) Statistical review of world energy 2014
Lu J (2009) Current situation and proposals for the development of natural gas industry in China. Nat Gas Ind 29(1):8–12
MLR (the Ministry of Land and Resources) (2014) 2014 China mineral resources (in Chinese)
NBS (National Bureau of Statics) (2011) China energy statistical yearbook 2011. China Statistical Press, Beijing
NBS (National Bureau of Statics) (2013) China energy statistical yearbook 2013. China Statistical Press, Beijing

Chapter 5
Overcapacity Challenges

5.1 Introduction

China has invested in building and continues to invest in building out infrastructure across industrial, transport and urban sectors. In 2013, China produced over half of the steel, cement and flat glass in the world and is the largest country in terms of energy production and consumption. As in any developing economy, there is a risk of inefficient investment in chasing growth targets often resulting in overcapacity. For China, this has become a serious issue. There is overcapacity today in several industrial sectors such as steel, cement, chemicals and in some locations power generation. The upshot has been to create stranded and underutilised assets that are often inefficient and have locked in early generation technologies with some creating serious environmental issues. Looking to the future, the question is how China can be smarter in making industrial investments.

As China looks to upgrade its economy away from low-cost manufacturing, the country cannot afford to continue to invest inefficiently or embed wasted energy in building new but underutilised industrial infrastructure. Failure to develop efficiently would undermine growth and long-term prospects for prosperity and competitiveness on the global stage. However, attaining a sustainable future will require China to change and upgrade its infrastructure, particularly in the power and refining sectors to address the country's chronic environmental conditions.

With increasing devolution of authority to the provinces; there is also an increasing need to consider China as a set of heterogeneous provinces, with remarkably different characteristics, rather than as a homogeneous country where one strategy fits all. Through recognising the differing characteristics and dynamic inter-play between provinces, it may be possible to pinpoint opportunities for optimisation of key networks such as mass transit, electricity generation, and gas transportation.

© Springer Science+Business Media Singapore 2016
Z. Li et al., *Informing Choices for Meeting China's Energy Challenges*,
DOI 10.1007/978-981-10-2353-8_5

In this chapter we illustrate initial work on the power and refining sectors to highlight the overcapacity issues. These are issues that foreign nations and companies have grappled with in the past.

5.2 Concepts and Reasons for Overcapacity

Overcapacity is normally regarded as a phenomenon created when the production capacity of an industry exceeds the actual demand from the market by a certain degree—resulting in underutilised, sub-optimal operation and even early retirement of assets and infrastructure. For example, according to the standard defined by the US Federal Reserve System, there is an overcapacity problem in a particular industry if its capacity utilisation rate is less than 80 %.

In China, serious overcapacity problems have developed in several energy and industrial sectors over many years. Taking oil refining and power generation as examples, shown in Fig. 5.1, the capacity utilisation rate of oil refining sector has remained well below 90 % in recent years, and there are localised overcapacity issues in the power generation sector, even when considering the need for spinning reserve in the power sector to meet peak demand, outages, and renewable intermittency. Similar overcapacity problems also exist in industrial sectors, such as iron and steel, cement, and flat glass. Of course industrial supply-demand mismatches are symptomatic of cyclical economic cycles, so overcapacity is the norm and unavoidable. However in China's planned economy there is no economic shake out of weaker inefficient players and overcapacity is allowed to persist and becomes exacerbated.

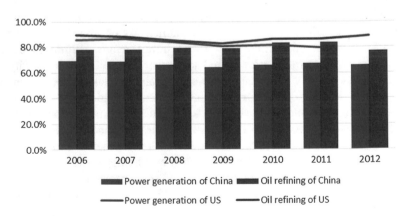

Fig. 5.1 Capacity utilisation rate of power generation and oil refining sector in China and US, 2006–2012

Overcapacity is already being felt in nascent industries such as coal conversion with China's coal-to-chemicals sector showing signs of over-heating (and more recently facing the collapse of product prices linked to crude). The China Petroleum and Chemical Industry Federation have warned that the situation may impact its sound progress in the long run. The amount of coal-to-oil projects, including those that are being designed or constructed, is totally over 25. For coal-to-olefin and coal-to-gas, there are over 60 projects in each category. It is not just the volume of projects that is a concern, but the huge challenges in the years ahead. The challenges include, but are not limited to, water supply, waste water treatment, pollutant disposal, and CO_2 emissions. Many of these issues may arise in tandem with profitability falling sharply from original forecasts and business plans.

Such overcapacity will not only reduce the efficiency of economic growth but also represent a significant misuse of energy in the building of an asset base and its associated infrastructure that then operates sub-optimally over a long productive life time. Low capacity utilisation rates or frequent instances of construction followed by early life shutdown, even demolition, of assets are regular occurrences across China.

Why has overcapacity appeared in China? The causes can be explained by the centrally planned approach that sees the government anticipating demand growth and determining the underpinning policy and planning to meet this growth.

The first cause is the changing market demand following rapid economic growth. After 2000, a dazzling economic boom created huge demand for production capacities in raw materials and energy within China. Rapid construction of production capacities was implemented to match the accelerating economic growth. However, since 2007, the rate of economic growth in China has slowed down; the contribution of export and capital investment to the economic growth also decreased. The rapid slowing of market demand growth led to overcapacity in sectors which continued to expand, especially in iron and steel, cement, chemicals, refining and power generation sectors, as the central plans were already embedded and being enacted. As a result, utilisation rates of assets in these sectors dropped to unsustainable levels (Fig. 5.2).

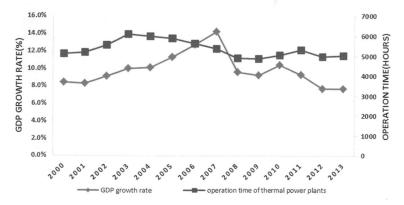

Fig. 5.2 China's GDP growth rate and the operation time of thermal power plants, 2000–2013 (NDRC 2009)

Table 5.1 China's four-trillion RMB investment plan (NDRC 2009)

Projects	Sums (billion RMB)
Build the low-rent houses and shanty towns transformation	400
Infrastructure construction in the rural area, such as water pipelines, power grids, road constructions, gas pipelines, and housing constructions	370
Major infrastructure construction such as railway, highway, airport, water conservancy facilities, and urban power grids	1500
Health care, education, and cultural undertakings	150
Ecological engineering, energy saving, and emission reduction	210
Innovation and adjustment of industrial structure	370
Restoration and reconstruction after catastrophes	1000

Nur Bekri, the chief of China's National Energy Administration recently indicated that energy rich provinces are seeking government help in selling their outputs. The NEA forecast that primary energy demand is expected to drop to 3.4 % p.a. through to 2020 and then 2.3 % p.a. until 2030. Hence in his view that "energy imbalances are urgent problems".

The second cause is the direct policy intervention in the construction of production capacities, including the intervention of investment policies and environmental policies. To stimulate economic growth after the world financial crisis, the Chinese Central Government introduced a four-trillion RMB investment plan in 2008 (shown in Table 5.1). Regional governments also formulated a series of their own supporting policies to stimulate economic growth. The stimuli of various investments are focused on bringing enormous financial capital into construction of production infrastructure rather than into that of consumption. This had the effect of exacerbating the imbalance between the booming construction of production capacities and the comparatively slow expanding consumption market, further intensifying China's overcapacity problems.

In addition, the financing strength of the SOEs at various levels from cities up to nation has resulted in irrational investment decisions due to the distortion caused by their "below market rate" financing costs. The improper decisions have led to rampant repetitive investments across different regions, which have in turn contributed to the mounting debt level in China. Currently, China's debt to annual GDP ratio has reached between 220 and 240 % of GDP in 2013. Debt repayment and default risks have become an onerous burden to the manufacturing sectors.

Moreover, as China has proactively sought to address local air pollution and greenhouse gas emissions, stricter environmental policies have inadvertently contributed to overcapacity, by over-promoting the development of cleaner technologies while limiting or forbidding others, swiftly creating stranded and worthless assets. Moreover, the flood of renewable power generation makes the competition across the power network more intense and creates overcapacity. All these factors reduce the utilisation rate of the new conventional, highly efficient coal-fired power generation, which is a phenomenon already witnessed in Europe.

The third cause is the lack of strategic planning and overall coordination on the construction of production capacities, especially at regional and sector levels. Although China has overarching five-year plans to manage the near-term development of its economy and society, longer term plans fail to receive the same level of attention with regional and sectorial planning particularly lacking high accuracy. The lack of integrated coordination at a regional level has resulted in repeated construction of similar production capacities in provinces and cities.

When a local government approves projects, there is limited consideration given to the regional optimisation of such assets, let alone improved integration with neighbouring provinces. Naturally each regional government pays more attention to its own near-term interests and GDP targets upon which their performance is measured than to long-term planning. Disorderly capacity expansion and overlapping investments have been allowed with "illegal new builds", where pressing demand has led to local authorities building and retrospectively gaining consent, undermining the role of central planning. These assets result in excessive competition and, with failure of newer investments being politically unthinkable, ultimately lead to serious overcapacity.

Our Analysis: In the Tsinghua-BP Clean Energy Centre, we recognised the mounting overcapacity problems in China and the lack of analysis performed in detailed long-term planning, especially at a regional and sectorial level. While the problem is across the economy, such as urban development and heavy industry, we are focused on the energy sector. We are developing tools and approaches based on solid understanding of technological limits and robust fundamentals to analyse this topic, including

(1) Regional Analyses: through decomposition of energy consumption into China's 30 regions to reveal regional disparities in economic growth and energy consumption.
(2) Oil Supply Chain: a dynamic model of the oil supply chain to optimise capacity and logistics.
(3) Power System Model: long term, regional optimisation of the power sector. Some of our initial analyses are discussed below.

5.3 Regional Analysis: Significant Disparities in Energy Consumption Patterns

Most previous studies on China's energy system have considered this huge country as a whole; our own regional work revealed large disparities in the energy-related characteristics of China's 30 provinces.

Based on a hybrid energy input–output model, the total energy consumption of different regions was decomposed and compared by the energy embodied

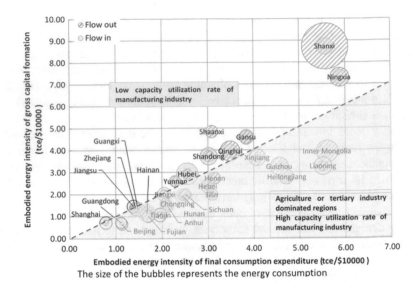

Fig. 5.3 Decomposed embodied energy intensity of 30 regions in China

in inter-regional trade, domestic capital investment, and domestic consumption expenditure. By carrying out the decomposed embodied energy intensity and economic development level, the 30 regions could be categorised into four distinct groups (shown in Figs. 5.3 and 5.4).

> Embodied energy is defined as the total (direct and indirect) energy required for the production of economic or environmental goods and services or the energy consumed in all activities necessary to support a process, including upstream processes. The embodied energy intensity is defined as the ratio of embodied energy and appropriate fund flows.

While the analysis is based on the latest available historic data, we conclude from our analysis that there are potential regional policy implications and priorities for better and more localised energy optimisation that reflect the characteristics of each of the four groupings, shown in different colours in Fig. 5.3.

For developed regions with low energy intensities, such as Shanghai, energy conservation should focus on promoting a low energy-consuming life style.

For underdeveloped regions with lower energy intensities, such as Guangxi, economic development is more urgent than energy conservation.

For developing and energy importing regions, improving energy efficiency in industry is necessary.

For developing and energy exporting regions, transforming primary energy into high value-added products would be beneficial for their economic development and energy conservation.

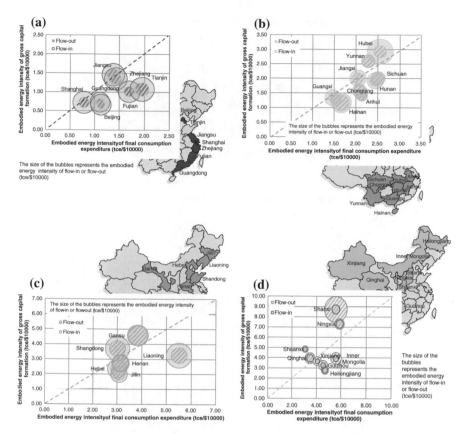

Fig. 5.4 Decomposed embodied energy intensity of the four provincial groupings (Li et al. 2014). **a** Group 1. **b** Group 2. **c** Group 3. **d** Group 4

5.4 Refining and Transport Fuel Sector—Meeting Cleaner Fuel Needs

In the March 9th 2015 edition of the China Daily, an article on the fight against pollution stated that the upgrading of vehicle emission standards had faced difficulties and postponement because of the low quality of the fuels available and disagreements inside government. The government aims to implement National Standard V by 2018 across the nation and provinces like Guangdong are seeking to accelerate its introduction. However there are commentators who have suggested that China should first do a thorough job implementing Euro IV standard and then leapfrog to Euro VI. This emphasises the investment challenges going forward in both the refining and vehicle sectors.

In refining and its associated supply logistics, the challenge involves transforming the industry from small refineries producing current fuels, to more advanced

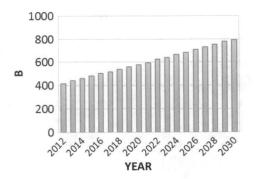

Fig. 5.5 The least required refining capacity under scenarios S1

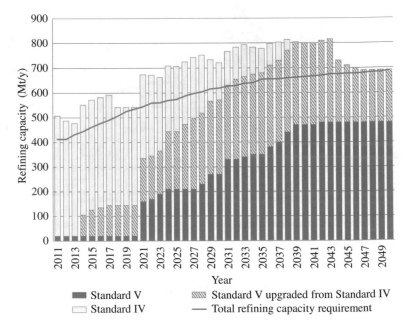

Fig. 5.6 Optimised route for refinery construction (Pan 2014)

plants manufacturing the cleaner burning fuels urgently needed to fight the pollution from vehicles and able to respond to the market in a more dynamic fashion.

The Clean Energy Centre has looked to model the development of such an oil supply chain to determine the optimum refining capacity, petroleum product storage, and fuel transport capability to manage both the uncertain future demand fluctuation (Fig. 5.5) and the evolving product quality specifications Fig. 5.6. There is much to be learnt from the similar refining industry transformations in the US and Europe that occurred from the 1970s until the present day.

5.5 Regional Model for Technology Optimisation of Power Generation Sector

Despite overcapacity at the country level, the regional disparities can create a very different local picture. For example in southern Hebei, the shortage can hit 3 GW during peak periods. Henan can see a 1.83 GW gap and Anhui a 1–2 GW deficit, while Jiangxi and Hunan will also witness small shortages. Hainan province, which is covered by China Southern Power Grid, has been experiencing shortages in 2013. More recently it has been revealed by the State Grid that in hydropower-rich Sichuan, the province has a total installed capacity of 68.6 GW, but peak load in the province is only 37 GW. Even with exports to other provinces the excess sits at just over 20 GW. The Sichuan Party Chief has identified that enhanced grid infrastructure is needed to export this excess of green power.

These examples indicate that there is significant opportunity for optimisation across the country by considering inter-regional optimisation. However this needs an appropriate tool to model power generation at a regional level and one that considers diurnal and seasonal demand patterns.

In our power generation model, we divided China into 10 regions (Fig. 5.7) and considered 10 power generation technologies, including pulverised coal-fired power generation (PC), PC with carbon capture and sequestration (PCC), integrated coal gasification combined cycle (IGCC), IGCC with carbon capture and sequestration (IGCCC), natural gas combine cycle (NGCC), nuclear power (NU), hydro power (HD), wind power (WD), biomass power (BM), and solar photovoltaic power (PV).

The model calculates an optimised technology pathway in 2011–2050 by minimising the total system cost across the entire forecast horizon for meeting demand against a backdrop of a variety of constraints. By considering future emissions and growing renewable intermittency, the work is also expected to demonstrate the considerable value of gas-fired power, which today remains limited and underappreciated by regulators (Fig. 5.8).

Fig. 5.7 10 regions considered in power generation model

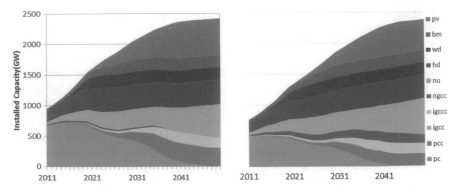

Fig. 5.8 The optimised technology pathway of power generation by regional model with single time block (*on the left*) and regional model with 12 time blocks seasonal load dispatch (*on the right*)

Inter-regional power transmission is also considered, and we further developed a load dispatch model considering the load change during each year and each day. Each year is divided into four loading period; Spring (March–May), Summer (June–August), Autumn (September–November), and Winter (December–February). Each day is divided into three load periods; low load period (0:00–8:00), medium load period (8:00–16:00), and high load period (16:00–24:00). Thus, the model can consider the various load curve and power utilisation curve, especially renewable power, in each region, and can optimise the distribution of power generation capacity by inter-region power transmission. The optimisation results above show that when daily load profile is considered, the system requires more natural gas-fired power which is almost absent from a more simple approach to meeting annual demand (Fig. 5.6). Other insights from the model imply inter-regional coordination of the construction of power generation capacity and peaking load sharing can optimise necessary power generation capacity. Such models can deliver better planning and coordination at a regional level and can relieve overcapacity.

Looking to our future work, we will model (a) the role of adopting a fully optimised regional approach to planning to deliver more efficient investment and reduce overcapacity and (b) the advantage of liberalised market mechanisms over regulated prices in delivering a more efficient allocation of capital. This will also need to consider three sensitivities to compare the total system cost and total build for each scenario.

(1) The single-region view of China's power market with fixed dispatch
(2) The multi-region view of China with fixed dispatch
(3) The multi-region view with optimised dispatch.

Personal View from Prof. Li Zheng, Director of Tsinghua-BP CEC Xi Jinping, in the 6th meeting of the central leading group in finance and economy, 13 June , 2014 stated that "Energy supply and consumption revolution should be adopted as a national strategy in the long run, focusing on (a) energy savings and total amount control of energy consumption, (b) developing a multiple-energy supply system, (c) upgrading the energy industry, (d) reintroducing market mechanism, (e) utilisation of international energy resources based on more international collaboration".

In my view this statement is a very important signpost for the future. China's economy growth used to feature scale, but severe economic efficiency and waste of energy also occurred in the course of huge amount of infrastructure building. Per the energy system, besides political and structural reasons, an important technical reason is lack of systematic and long-term planning, which depends on required modelling and planning ability. Due to China's large territory and huge difference between regions, building multiple-region and multiple-energy-source modelling and planning ability remains a challenge. This is one of the focuses of the Tsinghua-BP Phase II project. With the help of BP and similar companies, we can understand and learn more from existing international experiences. China's economy is entering a new normal state of slower growth. We need to contribute theoretical and technical support for decision making and planning in this new circumstance.

5.6 Conclusion

China's leadership has recognised the need to change the path of future economic development for China. This covers not only environmental challenges but also the type and quality of the investments to come.

In this chapter we have talked briefly about how industrial overcapacity has arisen in recent years. We believe that solutions cannot come from national target setting but require a more detailed region by region analysis, since there is such diversity and disparity across this large nation. Over the past 3 years we have developed our regional analysis and focused on the power sector in particular, creating a detailed power model. The power model can help to better understand how China can invest more efficiently in its power system focused on inter-regional optimisation and collaboration. We have also started to look at the refining sector where the current refining sector faces economic challenges yet has to invest to produce cleaner fuels.

Key messages are

Regional characteristics and dynamics need to be taken into account when thinking about strategy and planning in China.

Quality of capacity is as important a consideration as volume. Overcapacity has the effect of locking in too much technology of a certain vintage. The result is a less flexible and adaptive system that will incur further cost through either future refurbishment of existing assets or premature retirement followed by replacement new build.

China should learn from International institutions about government policy development, commercial and market models. Many European and North American institutions and companies have a proven track record in dealing with pollution, reconfiguring refineries, and cleaning up coal plants and dirty industries.

Our future work will look to focus on avoiding the worst causes of overcapacity and delivering environmental benefits and use the latest available data to model future scenarios.

References

Li Z, Pan LY, Fu F, Liu P, Ma LW, Angelo A (2014) China's regional disparities in energy consumption: an input–output analysis. Energy 78:426–438

National Development and Reform Commission (NDRC), Beijing, 2009. http://www.gov.cn/gzdt/2009-03/06/content_1252229.htm

Pan LY (2014) Dynamic modeling and embodied energy assessment of oil supply chain. Doctor dissertation, Tsinghua University

U.S. Energy Information Administration (EIA) Refinery utilization and capacity, http://www.eia.gov/dnav/pet/pet_pnp_unc_dcu_nus_m.htm

Chapter 6
Role of State-of-the-Art Technology

Policy and market structures provide a framework for investment and business development and can stimulate innovation in business models. Policy can also set aspirational targets for sustainable development. This is very evident in the energy sector. However, without cost effective, efficient, reliable, and environmentally benign technologies such aspirations cannot be met. Understanding what technologies can do and are available today is essential. How technologies may improve or new ones emerge is critical to modelling the future.

In this chapter we provide a brief overview of the current status of power generation technologies. Technical performance (capacity, efficiency) and economic performance of each technology are introduced, as well as relevant environmental performance. The power technologies include pulverised coal (PC), natural gas combined cycle (NGCC), nuclear, hydro, wind, biomass, and solar photovoltaic (PV).

6.1 Current Status of Power Technologies

6.1.1 Pulverised Coal (PC)

6.1.1.1 Technical Performance

A. Capacity

By the end of 2013, China's total installed capacity of power plants was 1258 GW, and thermal power occupied 69.2 % of the total national power capacity, i.e. 870 GW (see Fig. 6.1) (CEPP 2014). Among the thermal power units, 91.4 % of them were coal-fired power units. Except for the Huaneng Tianjin IGCC (integrated gasification combined cycle) power plants, all coal-fired power units were pulverised coal (PC) power units. By the end of 2014, China had operated

© Springer Science+Business Media Singapore 2016
Z. Li et al., *Informing Choices for Meeting China's Energy Challenges*,
DOI 10.1007/978-981-10-2353-8_6

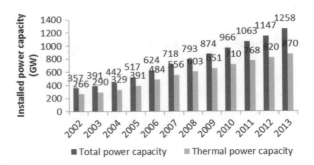

Fig. 6.1 Installed power capacity in China (2002–2013). *Source* CEPP, 2002–2013. China Electric Power Yearbook

almost one hundred 1000 MW ultra-supercritical PC units, becoming the largest ultra-supercritical PC market in the world.

B. Efficiency

The design thermal efficiency and coal consumption of China's typical condensing PC units are shown in Table 6.1. The steam quality (temperature, pressure) is observably improved from 300 MW subcritical PC (sub-PC), 600 MW supercritical PC (SPC), to 1000 MW ultra-supercritical PC (USPC). For 600 MW PC units, the gross efficiency (LHV) increases from 41.6 % of sub-PC, to 43.6 % of SPC, to 45.4 % of USPC. China's average net coal consumption of PC decreased dramatically from 383 gce/kWh in 2002 to 321 gce/kWh in 2013 (see Fig. 6.2).

6.1.1.2 Economic Performance

China's new-build PC units are mainly 300 MW subcritical CHP (combined heat and power), 600 MW SPC and 1000 MW USPC. The capital cost of PC units

Table 6.1 Design parameters of China's typical condensing PC units

PC units	Steam parameters		Gross efficiency (LHV) (%)	Gross coal consumption (gce/kWh)	Auxiliary power consumption rate (%)	Net coal consumption (gce/kWh)
	Temperature (°C)	Pressure (MPa)				
300 MW sub-PC	538/538	16.67	41.3	298	6.7	319.9
600 MW sub-PC	538/538	16.67	41.6	296	6.2–6.5	315.6–316.6
600 MW SPC	566/566	24.2	43.6	282	6.2–6.5	300.6–301.6
600 MW USPC	600/600	25	45.4	271	6–6.2	288.3–288.9
1000 MW USPC	600/600	27	45.7	269	5–5.5	283.2–284.7

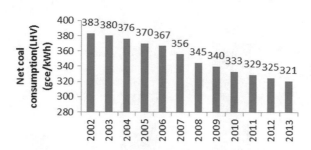

Fig. 6.2   Average net coal consumption of PC (gce/kWh)

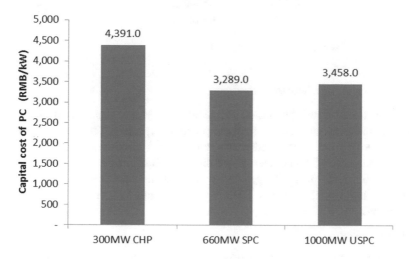

Fig. 6.3   Capital cost of new-build PC as of 2013 (RMB/kW)

(including PM (particulate matter) control, desulfurization, denitration (deNOx), and other environmental protection facilities) lies between 3200 RMB/kW and 4400 RMB/kW as of 2013 (CPECC 2014) (Fig. 6.3).

The levelised cost of electricity (LCOE) of PC varies with many factors. Figure 6.4 shows LCOE of PC units as of 2013 with the following assumptions: excluding value-added tax (VAT), 5000 h of the equivalent annual full-loading working hours, 6.55 % loan interest, 750 RMB/tce of coal price (including tax).

6.1.1.3   Environmental Performance

China's new-build USPC units are environmentally friendlier than the previous generation of coal power plants. For example, the air pollution emissions of Huaneng Yuhuan 1000 MW USPC units are only 0.134 g/kWh of SO_2 emissions, 1.09 g/kWh of NOx emissions and 0.114 g/kWh of PM emissions. In the following section, we discuss some of the emission management technologies.

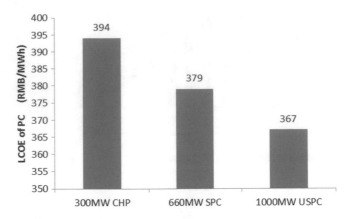

Fig. 6.4 LCOE of new-build PC as of 2013 (RMB/MWh)

A. PM—Particulate Matter

After 30 years of development, all PC plants with a unit capacity of 300 MW or above are installed with an electrostatic precipitator, a bag-filtering precipitator or an electrostatic-fabric integrated precipitator. Total PM emissions of power plants in 2011 were 1.55 million tonnes, decreasing 56.9 % from that of 2005 (3.6 million tonnes), while total electricity generation in 2011 increased by nearly 91 % above the output in 2005. The specific PM emissions declined from 1.33 g/kWh in 2005 to 0.4 g/kWh in 2011.

B. SO_2 emissions

By the end of 2012, flue gas desulfurization devices had been put into operation on more than 680 GW of PC (excluding CFB), accounting for about 90 % of PC capacity, to limit SO_2 emissions of PC within the national standards. In 2011, total SO_2 emissions of PC were 9.13 million tonnes, accounting for 41.2 % of China's total SO_2 emissions, and the specific SO_2 emissions of PC dropped from 6.4 g/kWh in 2005 to 2.3 g/kWh in 2011.

C. NOx

In 2011, total NOx emissions of PC were 10.73 million tonnes. With the implementation of "air pollutant emission standards of thermal power plants" (GB13223-2011) in 2012, more flue gas denitration devices have been deployed on China's PC plants. By the end of 2012, the flue gas denitration devices had been placed into operation on 230 GW of PC units, and were being installed or to be installed on more than 140 GW of further capacity. More than 98 % of denitration devices adopted the selective catalytic reduction (SCR) technology.

6.1.2 Gas Power

6.1.2.1 Technical Performance

A. Capacity

By the end of 2013, more than 150 gas power plants were deployed in China. NGCCs accounted for 3.4 % of the total power capacity in China, i.e. 43 GW, but generated only 2.2 % of the total power generation, i.e. 116 TWh (Figs. 6.5 and 6.6).

B. AEfficiency

By the end of 2012, there were 132 large-scale NGCC units (F class, E class,) with installed capacity of 36.96 GW, accounting for more than 90 % of total installed capacity of NGCC.

The configurations of NGCC systems are mainly "One gas turbine plus one HRSG" or "Two gas turbines plus one HRSG". F class (250 MW level) and E class (125 MW level) of gas turbines are widely used in China. The output of F class gas turbine is from 250 to 300 MW with the simple-cycle gross efficiency (LHV) of 36–39 %. The output of E class gas turbine is from 100 to 140 MW with the simple-cycle gross efficiency (LHV) of 32–34 %. Therefore in combined-cycle configurations, an NGCC with E class gas turbines can achieve 50–54 % of gross efficiency (LHV) and an NGCC with F class gas turbines can achieve 55–58 % of gross efficiency (LHV) (see Table 6.2). Meanwhile, the auxiliary power consumption rate of NGCC is very low, between 2 and 3 %.

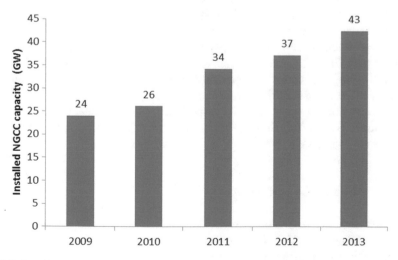

Fig. 6.5 Installed capacity of NGCC in China (2009–2013). *Source* China Electricity Council, http://www.cec.org.cn/

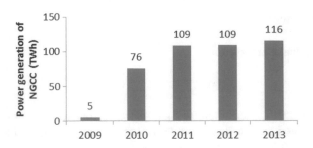

Fig. 6.6 Power generation of NGCC in China (2009–2013). *Source* China Electricity Council, http://www.cec.org.cn/

Table 6.2 Technical performance of gas turbine

		E class	F class
Gas turbine single cycle	Output (MW)	100–140	250–300
	Gross efficiency (LHV) (%)	32–34 %	36–39 %
NGCC	Output of "One gas turbine plus one HRSG" (MW)	~200	~380
	Output of "Two gas turbines plus one HRSG" (MW)	~380	~750
	Gross efficiency (LHV) (%)	50–54 %	55–58 %

6.1.2.2 Economic Performance

Today China can manufacture gas turbines and NGCC systems based on cooperation with foreign companies, such as GE, MHI, and Siemens. The investment cost of NGCC in China is about 3000 RMB/kW, at the same level as foreign companies. However, the annual maintenance cost of NGCC is expensive, about 30 $/kW, because China has yet to master the core technologies of gas turbines and the maintenance of gas turbines are still dependent on foreign companies.

The high levelised cost of electricity (LCOE), of which the fuel cost accounts for more than 70 %, is a major barrier to the development of NGCC in China. In 2013, the LCOE of F class "One gas turbine plus one HRSG" NGCC system was estimated to be 774 RMB/MWh with the following assumptions (CPECC 2014): 20 years operation, 15 years depreciation period; 3500 h equivalent annual full-load running time; 212 Nm^3/MWh of gas consumption for gas with 32,720 kJ/Nm^3 LHV (lower heating value); 2.67 RMB/Nm^3 gas price with tax.

6.1.2.3 Environmental Performance

Compared with PC technology, NGCC can effectively reduce the emissions of CO_2, SO_2, NO_X, and PM, because of the lower carbon content of natural gas and near zero levels of sulphur and ash (see Table 6.3). NOx emissions are produced by the combustion process itself (from the nitrogen content in air) but NGCC technologies incorporate low NOx burners and SCR if needed.

Table 6.3 Emissions of PC and NGCC (unit size: 500 MW)

Technology	SO_2/(t a^{-1})	NOx/ (g kWh^{-1})	CO_2/(g kWh^{-1})	PM/ (g kWh^{-1})	Slag/ (g kWh^{-1})
PC	1.37–2.74	0.422–2.055	731	0.034–0.103	46.8
NGCC	0.002	0.274–1.256	411		0

Source NETL. Cost and Performance Baseline for Fossil Energy Plants Volume 1: Bituminous Coal and Natural Gas to Electricity. Revision 2, November 2010, DOE/NETL-2010/1397

6.1.3 Nuclear

6.1.3.1 Technical Performance

By the end of September, 2014, 21 nuclear power units were placed into operation in China, and the total capacity was 19 GW (see Table 6.4).

Pressurised water reactor (PWR) technologies, including AES-91, M310, CPR1000, AP1000, EPR, are either in-service or being-built in China, as well as one heavy water reactor technology (CANDU6), and one high-temperature gas-cooled reactor (HTGR) in Shandong province.

The CPR1000 design, improved from the 2nd generation of nuclear power technology M310, is used for most of China's nuclear power units. Domestic companies can manufacture most components of the nuclear and conventional islands.

AP1000 and EPR are two 3rd generation nuclear power technologies to be constructed. The AP1000, developed by Westinghouse Electric Corporation of US, will be used in two units of the Zhejiang Sanmen Nuclear Power Plant and two units of Shandong Haiyang Nuclear Power Plant. EPR is developed by the Areva Corporation of France. Two EPR units with capacity of 1.75 GW, the largest nuclear power unit in China, are being built in Guangdong Taishan Nuclear Power Plant facility.

Table 6.4 In-service nuclear power plants in China by September, 2014

Nuclear power plants	Technology	Output (MW)
1st stage of Qinshan	Pressurised water reactor (PWR)	1 × 320
2nd stage of Qinshan	PWR (CNP650)	4 × 650
3rd stage of Qinshan	Heavy water reactor (CANDU6)	2 × 728
Dayawan	PWR (M310)	2 × 984
1st stage of Lingao	PWR (CPR1000)	2 × 990
1st stage of Tianwan	PWR (VVER)	2 × 1060
2nd stage of Lingao	PWR (CPR1000)	2 × 1080
Fuqing	PWR (CPR1000)	1 × 1080
Ningde	PWR (CPR1000)	2 × 1080
Hongyanhe	PWR (CPR1000)	2 × 1080
Yangjiang	PWR (CPR1000)	1 × 1080

One 200 MW HTGR unit is under construction in Rongcheng, Shangdong Povince. Tsinghua University is the supplier of the reactor technology.

6.1.3.2 Economic Performance

The capital cost of in-service nuclear power units (the 2nd generation or improved 2nd generation technologies) is about 13,000 RMB/kW. For the 3rd generation nuclear power units, the capital cost of AP1000 units is between 14,000 and 15,000 RMB/kW, and that of EPR units is within 17,000 RMB/kW.

In 2013 July, the National Development and Reform Commission (NDRC) issued the 'Notice on relevant issues of pricing mechanism of nuclear power', which changed the grid electricity price from "individual price" to "benchmark price", i.e. 0.43 RMB/kWh, for nuclear power units deployed after 1 January 2014. In the conditions, 9 % of the internal rate of return (IRR) can be achieved by the No. 1 unit of Hongyanhe Power Plant.

6.1.4 Renewable Power

6.1.4.1 Technical Performance

Hydropower (HD)
China's hydropower capacity and annual power supply surpassed the United States and Canada, respectively, in 2004 and 2005, ranking first in the world. At the end of 2013, total installed capacity of hydropower was 280 GW, accounting for 22.3 % of China's total power capacity (see Fig. 6.7).

Wind Power
In 2012, there were 7872 new-build wind power units and the installed capacity was 13 GW in China; A total of 53,764 wind power units have been installed across China with cumulative installed capacity of 75 GW (Fig. 6.8). By the end of 2012, the total offshore wind power capacity had reached 390 MW, including 262 MW of intertidal wind power and 128 MW of offshore wind power. The unit capacity of domestic wind power was 1.5–2 MW in 2008, and the largest wind power unit deployed was 6 MW in 2012. By the end of 2014, China's accumulative installed wind power capacity reached 96 GW.

Biomass Power
Biomass is usually used for power generation via direct combustion, co-firing, and gasification. By the end of 2013, China's biomass power capacity was just under 8.7 GW, of which straw, bagasse, and wood power constituted about 5.1 GW, and municipal solid waste power was a further 3.6 GW. Biomass power output was 38.3 TWh, of which straw, bagasse, and wood power output was 20.7 TWh, and waste to power output was about 17.6 TWh (Fig. 6.9).

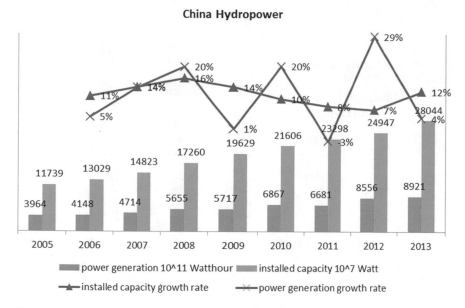

Fig. 6.7   Installed capacity of hydro power in China (2005–2013)

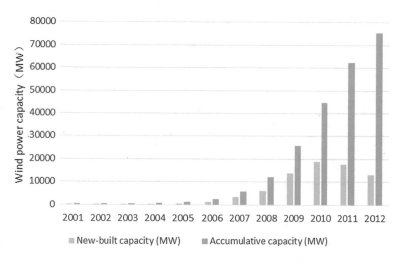

Fig. 6.8   China's wind power capacity(2001–2012)

The unit capacity, net efficiency (LHV) and specific capital cost of different biomass technologies are shown in Table 6.5.

Solar Photovoltaic (PV)

In recent years, solar PV technology has been developing rapidly in China. Installed capacity of PV was only 20 MWp in 2007. By the end of 2012, China's total installed capacity of PV was about 7 GWp, 36.4 % of which was distributed PV.

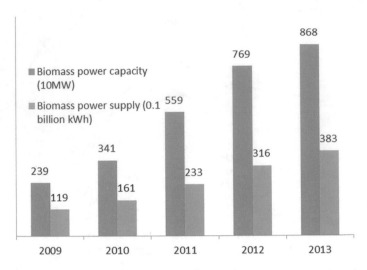

Fig. 6.9 Capacity and output of biomass power (2009–2013). *Source* Statistical yearbook of the power industry, China Electricity Council

Table 6.5 Typical technical and economic performance of biomass power

Technology	Unit capacity	Net efficiency (LHV)	Specific capital cost
Direct-fired	10–100 MW	20–40 %	1975–3085 $/kW
CHP	0.1–1 MW 1–50 MW	60–90 % thermal efficiency 80–100 % thermal efficiency	3333–4320 $/kW 3085–3700 $/kW
Co-firing	5–100 MW (existing) > 100 MW (new-build)	30–40 %	123–1235 $/kW + power plant cost
Landfill gas	<200 kW–2 MW	10–15 %	
BIGCC	5–10 MW demonstration 30–200 MW future	40–50 %	4320–6170 $/kW 1235–2740 $/kW

By the end of December 2013, China's cumulative installed capacity of PV reached 17 GWp, including 11 GWp of large-scale PV and 6 GWp of distributed PV.

Crystalline silicon and thin film are two dominant technologies. The crystalline silicon technology is matured, occupying more than 90 % market share. Monocrystalline and polycrystalline are two important versions of the crystalline silicon technology. The efficiency of monocrystalline silicon technology (up to 15 %, and expected to achieve 25–28 %) is higher than that of polycrystalline, but its cost is higher. The thin film technology is a new-developed technology. It uses fewer raw materials than the crystalline silicon technology and can offer building integrated solutions, but its efficiency is low (Table 6.6).

Table 6.6 Technology performance of PV technologies (2050).

	Crystalline silicon		Thin film		New concept	
	Monocrystalline	Polycrystalline	No-crystalline/less-crystalline thin film	Polycrystalline thin film	High performance	Low cost
Efficiency (%)	24–28 %	20–25 %	CIS:22–25 % Si: 20 %	6–8 %	>40 %	10–17 %
Lifetime (year)	40–50	40–50	30–35	30	>25	10–15
Market share	Segment	Mainstream	Mainstream	Mainstream	Segment/mainstream	Mainstream
Application	Limited by install area	All	All	Customer, building	Limited by install area	All

Source International Energy Agency, Energy Technology Perspectives 2008

6.1.4.2 Economic Performance

Hydropower (HD)

Hydropower is one of the low-cost power technologies, because hydropower units can run for 50–100 years and have no fuel costs. In developing countries, the capital cost of hydropower is no more than 1000 $/kW, and the LCOE of small hydropower usually is 20–60 $/MWh. The capital cost of China's large hydropower (more than 300 MW) is about 1538 $/kW, LCOE is 0.031–0.038 $/kWh; the capital cost of small hydropower (less than 50 MW) is approximately 923.07 $/kW, and LCOE is 0.038–0.043 $/kWh.

For large hydropower there are indirect consequential costs associated with moving communities and impacts on local ecology. These are not easily quantified and therefore are not included in the study.

Wind Power

75 % of the total costs of wind power are associated with capital, including wind power units, construction, electronics, and grid connection. Infrastructure construction costs typically account for about 10 % of the total investment of onshore wind power, while it rises up to 40 % for offshore wind power, which means the investment costs of offshore wind power are 50–100 % higher than that of onshore wind power.

A research report issued by CITIC Securities showed that the investment was 1230–1538 $/kW and LCOE was 0.066–0.081 $/kWh for a 50 MW wind power unit with assumptions of 30 % of self-own investment and 1800 h of equivalent full-loading working hours. National Development and Reform Commission (NDRC) had graded wind power benchmark prices in 2009: 0.078, 0.083, 0.089, and 0.094 $/kWh. Therefore, the LCOE of wind power was lower than the benchmark prices set by the NDRC.

Biomass Power

Because the biomass direct-fired boilers and pretreatment equipment introduced from abroad are expensive, the specific capital cost of biomass direct-fired power units is greater than that of PC. For a typical biomass combined heat and power (CHP) plant with two straw-fired boilers of 65 t/h (imported from abroad or foreign cooperatively manufactured) and one C25 extraction condensing turbine, the total capital cost is about $45 million, and the specific capital cost is 1790 $/kW, which is about 35 % higher than that of coal-fired CHP plants (1308 $/kW).

The LCOE of China's biomass direct-fired power is 0.108–0.123 $/kWh; LCOE of domestic small-scale biomass gasification power is 0.061–0.076 $/kWh; LCOE of waste incineration power is 0.108–0.123 $/kWh. The subsidy policy for biomass direct-fired power has been issued in China: the on-grid price of electricity for biomass direct-fired power is equal to coal power benchmark price plus 0.038 $/kWh.

Solar Photovoltaic (PV)

Capital cost of on-grid PV includes photovoltaic cells, grid-connected inverter, power measurement and cables, equipment transport, installation, and commissioning. By the end of 2012, specific investment of China's on-grid PV was 2.3 $/Wp.

6.1.5 Remarks

The performance of existing power technologies is shown in Table 6.7, where the integrated gasification combined cycle (IGCC) is not included, because only one IGCC power plant has been deployed up to now.

Pulverised coal (PC) has a high LHV efficiency (40–45 %) and low capital cost (3500–4500 RMB/kW) and LCOE (about 380–420 RMB/MWh), so PC is China's dominant power technology, accounting for half of China's coal consumption and producing more than 70 % of electricity output.

However, PC is also a major air pollutant producer, contributing most of China's CO_2, SO_x, NOx and PM emissions. Compared with PC, NGCC has a higher efficiency and lower emissions of CO_2, SOx, NOx, PM, and other pollutants. However, the LCOE of NGCC is higher than that of PC because of the higher fuel cost, though the capital cost of NGCC is less than that of PC. Therefore, NGCC is mostly employed tactically as a peak load power plant or as a combined heat and power (CHP) plant.

Nuclear power is a zero carbon and zero air pollutant power technology. The LCOE of nuclear power is not high because of its long annual working hours. However, the total capacity of nuclear power is relatively small, accounting for less than 2 % of national power capacity, due to the domestic limitation of uranium fuel resource and general safety concerns.

Hydropower is a clean renewable power technology. The LCOE of hydro power can be lower than that of PC, because of its long operational lifetime. Hydropower contributes more than 15 % of China's electricity output. Other renewable energy, such as wind, biomass and PV, can also generate clean power. The capacity of renewable energy has doubled and redoubled recently. The LCOE of renewable energy is higher than that of PC due to the shorter annual working hours (for wind power and solar PV) or higher fuel cost (for biomass).

6.2 Future Potential of Power Technologies

The development and deployment of energy technology is impacted by many factors such as technological maturity, economics, regulation (particularly local environmental protection), global concerns about the climate change, and societal acceptance. Looking forward future power technologies will need to be affordable, efficient, and clean. In particular, with the gradual but steady increase in environmental protection awareness, low pollutant emissions and low-carbon emissions will become the basic performance requirement of the future energy technology.

For China, the strategy will be to:
optimally develop PC
preferentially develop hydropower
safely and efficiently develop nuclear power
actively promote new power technologies
develop gas power appropriately (mainly NGCC).

Table 6.7 Performance of existing power technologies in China

Technology	PC	NGCC	Nuclear	Hydro power	Wind power	Biomass	Solar PV
Fuel	Coal	Natural gas	Uranium			Biomass	
Typical unit size (MW)	300–1000	200–780	1000		1.5–6	10–100	
Efficiency (%LHV)	40–45	50–58				20–40	20–28
Capacity as of 2012 (GW)	753.8	37.2	13	249	75.3	7.7	3.4
Capacity share as of 2012 (%)	65.7	3.2	1.1	21.8	6.5	0.6	0.3
Output as of 2012 (TWh)	3710.4	109.2	98.3	855.6	103	31.6	3.6
Output share as of 2012 (%)	74.4	2.2	2	17.2	2.1	0.6	0.1
Annual equivalent full-load working hours (hours)	5000–6000	3500	7500–7900	3000–3700	1800–2000		2000
Construction time (year)	3	2	5	5	1		
Lifetime (year)	20–30	20	50	50	20		40–50
Capital costs (RMB/kW)	3500–4500	3000	13,000	9000	7400–9400	12,000	14,000–30,000
LCOE (RMB/MWh)	380–420	715			400–500	650–740	500–1600
PM emission (g/kWh)	0.034–0.103						
CO_2 emission (g/kWh)	800–1100						
SOx emission (g/kWh)	2.3	0.002					
NOx emission (g/kWh)	0.422–2.055	0.274–1.256					

The 12th Five Year Plan of Power Industry states that: by 2020, national power capacity will reach 1935 GW, including 1170 GW of PC, 360 GW of hydropower, 60 GW of pumped storage power, 80 GW of nuclear power, 50 GW of gas power, 180 GW of wind power, 25 GW of solar power, and 10 GW of biomass, tidal, geothermal, and other renewable power. Therefore, the total capacity of non-fossil fuel power will reach 715 GW in 2020, accounting for 37 % of the national power capacity. Non-fossil fuel power can generate 2.3 trillion kWh, accounting for 27.3 % of the total national power generation. Non-fossil power can replace 7.3 billion tce of fossil energy, accounting for about 14.5 % of the primary energy consumption.

6.2.1 PC Trends

Efficiency improvement is still the main development route of PC and improving steam enthalpy is an effective way. Therefore, double reheat is being employed in supercritical PC units, and ultra-supercritical PC units with high-enthalpy steam (35 MPa, 700 °C) will be established in the future, since its efficiency can improve to as much as 50 %. Meanwhile, the pollutant control technology will be diversified and integrated, including dust precipitation, desulfurization, denitration, CO_2 capture, and in this way the pollutant emissions of PC can be reduced to the existing pollutant emission standards of gas power.

6.2.2 Gas Power Trends

Gas power is expected to be developed at a limited rate due to the high cost of the fuel. We expect to see large-scale gas power deployed to meet the demand of grid dispatch, and located at pipeline terminals and LNG import and re-gas facilities.

For gas turbine technology itself, China plans to develop G/H class gas turbines with higher inlet temperature (1500 °C) and higher efficiency, but China needs to first master the existing E/F class gas turbine technology before 2020.

6.2.3 Nuclear

Nuclear power technology is expected to become safer, more efficient and produce less radioactive waste. Passive safety technology and serious accident prevention designs will be adopted by the 3rd generation pressurised water reactor (PWR) technologies. Looking further out 20 more years, new waste mitigation technologies, alternative nuclear fuel cycles and safer subcritical designs are expected in 4th generation nuclear power technology.

Future nuclear technology will basically develop in three steps, from thermal neutron reactors today to some fast breeder reactors and towards controlled fusion reactors. High-temperature gas-cooled reactor technology and the fast breeder reactor technology will be developed independently. The fusion reactor technology will be researched via international cooperation, such as the International Thermonuclear Experimental Reactor (ITER) Plan.

China will likely be the proving bed for new nuclear technologies but nuclear technology development takes considerable time and resources, counted in decades and billions.

6.2.4 Renewable Power

Renewable power will be actively developed by China with full consideration of social affordability of electricity prices and sustainability of international competitiveness.

Hydropower

Large- and small-scale hydropower will continue to be developed in China. In addition, pumped storage power will also be constructed to help manage intermittent renewable power and improve the flexibility of the power system. High-efficiency, large-scale hydropower turbines will be utilised for both these applications (700 and 450 MW respectively). In the near term (5 years), 69 % of the recoverable hydroelectric resource will be developed, i.e. the hydro resource in eastern and central China will be fully exploited and 63 % of hydroelectric resource in western China will be deployed.

Wind Power

Wind power technologies will continue to be deployed across China at various scales both onshore and offshore. Wind power turbines are expected to increase in size with machines as large as 10 MW in the next 5 years. However, curtailment of wind power is common in China to meet the grid dispatch. Therefore, energy storage technologies, such as electricity storage, heat storage, and pumped storage, will be needed to accommodate curtailed wind power.

Biomass

Co-firing technologies can compete with PC, if the current preferential policies and subsidies for direct-fired technologies were established for them. Therefore, co-firing technologies will develop rapidly in the future. The capacity of co-firing power units can approach 600 MW. In addition, garbage power capacity is expected to double and dispose of millions of tonnes of garbage annually.

PV

The acquisition, storage, and utilisation of solar energy will be integrated in the future PV system. Unit PV inverter capacity can reach 1 MW, and the automatic sun-tracking device will be widely applied. PV is the next point of power growth of China, and installation of PV will run at a high speed in the next decade. It is expected that more and more distributed PV will be established for the smart grid.

6.3 Conclusions

We have focused on power generation technologies in this chapter. We have summarised the current state of these technologies. Although coal-fired power generation dominates, China has made significant strides in deploying the world's most efficient coal power plants. With added clean-up technologies, these coal power plants can reduce emissions of air pollutants. These are urgently required to mitigate the smog issue across the country.

Natural gas combined-cycle plants can offer China cleaner fossil fuel power and reduce CO_2 emissions relative to coal. However, the lack of plentiful cheap natural gas constrains its growth in the near term.

China is the world's biggest renewable market. It has exploited its large hydroelectric resource and there is still space to develop more, although this needs to be mindful of ecological and community impacts. For wind and solar PV, China continues to expand its renewable base rapidly and both technologies have space to improve in terms of capital cost and conversion efficiency. By contrast, biomass offers only niche opportunities and therefore will provide only localised solutions.

China also has the world's largest nuclear construction programme. The scale and cost of nuclear installations lead to a slower growth trajectory than that of coal and non-hydro renewable.

Looking to the future, only solar PV, nuclear, or energy storage technologies could offer breakthrough potential but these are unlikely in the next 10–15 years. Combustion plants will offer incremental improvements but it will be emissions reduction technologies that will shape their future, particularly if carbon capture is required.

One final technological area of importance is the improvement of energy efficiency in the consuming end of the value chain. There is a big prize in simply reducing the demand for energy and electricity. China must also pursue this as it pursues economic advancement in the decades ahead.

References

CEPP (China Electric Power Press) (2014) China Electric Power Yearbook 2013. Beijing, 2013
CPECC (China Power Engineering Consulting Group Corporation) (2014). Reference cost index for the quota of thermal power engineering design (2013). Beijing

Chapter 7
Modelling China's Future Energy Infrastructure: An Introduction to the Modelling Methodology

In this chapter, we introduce our energy systems modelling methodology. We have developed and used a multi-regional superstructure optimisation model that can optimise the regional development of China's power sector. This work includes the planning of energy redistribution through regional and national power transmission, as well as a better spatial utilisation of renewable resources with due consideration to regional distributions, resources, and demand.

7.1 Introduction

Power and transport are amongst top energy-consuming energy sectors in China. Previously, there have been some studies on energy consumption in the two sectors and projections to their future trends. However, most existing studies take China as a single entity with the same boundary conditions and scenario assumptions, such as gross domestic product (GDP) growth rate, energy (power, fuel) demand per capita, and others. Regional differences are not properly considered in these studies. Therefore, although outcomes of these studies could provide some useful insights at a national level, they are not able to address unique challenges in different regions.

7.2 Methodology Description

7.2.1 Power Model Structure and Assumptions

In our power sector model, China is modelled as 10 regions reflecting the current physical structure of regional transmission networks. As shown in Fig. 7.1, the 10 regions are as follows: Northeast (Heilongjiang, Jilin, Liaoning, and East Inner Mongolia),

© Springer Science+Business Media Singapore 2016
Z. Li et al., *Informing Choices for Meeting China's Energy Challenges*,
DOI 10.1007/978-981-10-2353-8_7

Fig. 7.1 Regional division
of China

North (Beijing, Tianjin, Hebei, Shanxi, and West Inner Mongolia), Shandong, East (Shanghai, Jiangsu, Zhejiang, and Anhui), Fujian, South (Guangdong, Yunnan, Guizhou, and Guangxi), Chuanyu (Sichuan and Chongqing), Central (Jiangxi, Hubei, Hunan, and Henan), Northwest (Shaanxi, Gansu, Ningxia, and Qinghai) and Xinjiang. Hainan, Tibet, Hong Kong, Macau, and Taiwan are excluded as these regions have relatively independent grids and small regional power demands.

Ten types of power generation technologies are considered: pulverized coal (PC), PC with carbon capture and storage (CCS) (PCC), integrated gasification combined cycle (IGCC) (China's first 250 MW IGCC Demonstration Power Project was put into operation in Huaneng Tianjin Gasification Co., Ltd. in 2012), IGCC with carbon capture and storage (IGCCC), natural gas combined cycle (NGCC), nuclear (NU), hydropower (HD), wind power (WD), biomass power (BM), and photovoltaic (PV). These technologies are classified into three categories. The first category of plant, which includes NGCC, NU, HD, WD, BM, and PV, is decommissioned only at the end of its expected lifetime. The second category of plant, which includes PC and IGCC, may be retrofitted with CCS or decommissioned before reaching its expected lifetime to reduce carbon emissions. The third category, which includes PCC and IGCCC, is constructed directly or from retrofitting. Where investment decisions are possible, the model makes the optimal decision.

This multi-region model is capable of reflecting inter-region power transmission line capacity, transmission losses, and use of system operational costs. Therefore, the delivery of power to meet demand must take transmission losses into account when considering the amount of output required from power plants. Consequently, introducing the inter-region power transmission reflects and facilitates a more reasonable allocation of power generation and integration of resources in meeting demand. In this study, the international exchange is neglected due to the little amount of power transmission with the neighbouring countries. In 2013, the amount of import and export electricity is 7.44 and 19.67 TWh, respectively, accounting for 0.14 and 0.36 %, respectively, of the power demand (5420 TWh).

CCS technology can capture 90 % of the CO_2 emitted by coal-fired power plants, and the development of CCS can be driven largely by carbon reduction policies. In this study, a long-term cap-and-trade policy for carbon emission was modelled to reflect a future carbon emission reduction policy instrument. Each region is allocated a cap on CO_2 emissions and a CO_2 price. The cap and price varies across each year of the planning horizon. The policy acts as both a penalty in proportion to the regional CO_2 price to be paid by the regional power sector where regional emissions exceed the regional cap; and a benefit where the actual regional emissions are lower than the regional cap.

7.2.2 Mathematical Equations

The mathematical formulation of the multi-region model is presented in this section. To provide a better understanding of the equations, the physical meanings of parameters and variables in the model are shown in Appendix A. Four sets, including t, r, g, and f, stand for year, region, type of power plant, and fuel type, respectively. In the equations, year t' and t'', sharing the same set, are used to distinguish from the year t, while r' shares the same set as r.

7.2.2.1 Objective Function

The objective function of the model is to minimize the accumulated total cost over the planning period from 2011 to 2050 of meeting electricity demand. The accumulated total cost is the sum of all total costs for each region and each year, as expressed in Eq. (7.1).

$$\text{atc} = \sum_{t=2010}^{2050} \frac{\sum_r \text{rc}_{r,t}}{(1+I)^{(t-2010)}} \tag{7.1}$$

The total cost of each year consists of six different types of cost: capital cost for construction, capital cost for retrofit, operation and maintenance cost, fuel cost, CO_2 emission cost and transmission cost. The total power sector cost of a given region r in year t can be calculated as follows:

$$\text{rc}_{t,r} = \text{tinv}_{t,r}^{nb} + \text{tinv}_{t,r}^{rf} + \text{tom}_{t,r} + \text{tfc}_{t,r} + \text{cec}_{t,r} + \text{ptrc}_{t,r} \tag{7.2}$$

7.2.2.2 Capital Cost for Construction

For the first and third categories of power plants, the capital cost is levelised to each year of the expected lifetime. The capital cost for constructing these types of power plants can be calculated as follows:

$$\text{inv}_{g,t,r}^{nb} = \sum_{t'=t-T_g+1}^{t} \left(\text{CAP}_{g,t'}^{nb} \cdot nb_{g,t',r} \cdot \frac{I}{(1+I) \cdot \left(1 - (1+I)^{-T_g}\right)} \right) \quad (7.3)$$

For the second-category power plants, the actual lifetime may be shorter than the expected lifetime if the optimal decision is to retire a plant earlier. The capital cost for constructing these types of plants consists of three parts: capital cost for plants retired early, retrofitted plants,and plants retired at the end of expected lifetime. This variable, along with its three components, can be expressed as follows:

$$\text{inv}_{g,t,r}^{nb} = \text{inv}_{g,t,r}^{nb,er} + \text{inv}_{g,t,r}^{nb,rf} + \text{inv}_{g,t,r}^{nb,nr}$$

$$= \sum_{t'=t-T_g+1}^{t} \left[\begin{array}{l} \displaystyle\sum_{t''=t+1}^{t'+T_g-1} \left(\text{CAP}_{g,t'}^{nb} \cdot er_{g,t',r}^{t''} \cdot \frac{I \cdot (1+I)^{-1}}{1 - (1+I)^{-(t''-t')}} \right) \\[3mm] + \displaystyle\sum_{t''=t'+1}^{t'+T_g-1} \left(\text{CAP}_{g,t'}^{n,b} \cdot rf_{g,t',r}^{t''} \cdot \frac{I \cdot (1+I)^{-1}}{1 - (1+I)^{-T_g}} \right) \\[3mm] + \text{CAP}_{g,t'}^{n,b} \cdot \left(nb_{g,t',r} - \displaystyle\sum_{t''=t'+1}^{t'+T_g-1} \left(rf_{g,t',r}^{t''} + er_{g,t',r}^{t''} \right) \cdot \frac{I \cdot (1+I)^{-1}}{1 - (1+I)^{-T_g}} \right) \end{array} \right]$$

$$(7.4)$$

Total capital cost for construction of the power sector in region r in year t was calculated as follows:

$$\text{tinv}_{t,r}^{n,b} = \sum_{g} \text{inv}_{g,t,r}^{nb} \quad (7.5)$$

7.2.2.3 Capital Cost for Retrofit

For power plants in the second category, the capital cost for retrofit is levelised over the period after retrofitting. The discounted capital cost in year t for retrofitting the plants constructed in year t and retrofitted in year t'' can be calculated as follows:

$$\text{inv}_{g,t,r}^{rf} = \sum_{t'=t-T_g+1}^{t-1} \sum_{t''=t'+1}^{t} \left(rf_{g,t',r}^{t''} \cdot \text{CAP}_{g,t'}^{rf} \cdot \frac{I}{(1+I) \cdot \left(1 - (1+I)^{-(t'+T_g-t'')}\right)} \right)$$

$$(7.6)$$

The total capital cost for retrofit of a power sector in region r in year t can be calculated as follows:

$$\text{tinv}_{t,r}^{rf} = \sum_g \text{inv}_{g,t,r}^{rf} \tag{7.7}$$

7.2.2.4 Operation and Maintenance Cost

The total operation and maintenance cost of a power sector in region r in year t can be calculated as follows:

$$\text{tom}_{t,r} = \sum_g \text{om}_{g,t,r} = \sum_g \left(\mu_{g,t} \cdot ic_{g,t,r} \right) \tag{7.8}$$

7.2.2.5 Fuel Cost

Fuel cost is the product of fuel price and total fuel consumed. The total fuel cost of a power sector in region r in year t can be calculated as follows:

$$\text{tfc}_{t,r} = \sum_f fc_{f,t,r} = \sum_f \left(\text{FP}_{f,t,r} \cdot \text{tfd}_{f,t,r} \right) = \sum_f \left(\text{FP}_{f,t,r} \cdot \sum_g fd_{f,t,r}^g \right)$$
$$= \sum_f \left(\text{FP}_{f,t,r} \cdot \sum_g \left(pg_{g,t,r} \cdot \text{FCR}_{f,g,t} \right) \right) \tag{7.9}$$

7.2.2.6 CO_2 Emission Cost (or Revenue)

CCS was assumed capable of capturing 90 % of emitted CO_2. The CO_2 emissions are the product of fuel consumption and CO_2 emission intensity. The total CO_2 emission of a power sector in region r in year t can be calculated as follows:

$$\text{tce}_{t,r} = \left(fd_{\text{coal},t,r}^{pc} + 0.1 fd_{\text{coal},t,r}^{pcc} + fd_{\text{coal},t,r}^{igcc} + 0.1 fd_{\text{coal},t,r}^{igccc} \right) \cdot \text{CEI}_{\text{coal}}$$
$$+ fd_{ng,t,r}^{ng} \cdot \text{CEI}_{ng} + fd_{u,t,r}^{nu} \cdot \text{CEI}_u \tag{7.10}$$

According to the cap-and-trade policy modelled, a CO_2 cap and a CO_2 price are assumed for each region and each year. If the actual CO_2 emissions surpass the CO_2 cap, a CO_2 emission cost is paid according to the excess amount and CO_2 price. On the contrary, if the actual CO_2 emissions are lower than the CO_2 cap, CO_2 emission revenue is acquired according to CO_2 price and the emission reductions relative to the cap. The total CO_2 emission cost (or revenue) of a power sector in region r in year t can be calculated as follows:

$$\text{cec}_{t,r} = \text{CP}_t \cdot \left(\text{tce}_{t,r} - \text{CECAP}_{t,r} \right) \tag{7.11}$$

7.2.2.7 Transmission Cost

The region importing power is assumed to bear the transmission costs. The total transmission cost of importing power to a region r in year t can be calculated as follows:

$$\text{ptrc}_{t,r} = \begin{cases} 0 \\ \sum_{r'} \text{TRC}_{r,r'} \times \text{tcrr}_{r,r',t} \end{cases} \tag{7.12}$$

7.2.3 Constraints and Conditions

7.2.3.1 Power Demand and Transmission

Power demand in region r in year t in the multi-region model is considered equal to the algebraic sum of the power generation and the transmission of the corresponding region and year. Thus, power demand can be calculated as follows:

$$\text{PD}_{r,t} = \sum_g pg_{r,g,t} + tc_{t,r} = \sum_g ic_{r,g,t} \cdot \text{OH}_g + tc_{t,r} \tag{7.13}$$

The total power transmitted into region r in year t is equal to the algebraic sum of actual power transmitted from all the other regions to region r in year t, which can be expressed as follows:

$$tc_{t,r} = \sum_{r'} \text{tcrr}_{t,r,r'} \tag{7.14}$$

If region r is the exporting region, the power transmitted from region r' to region r should be negative and equal to the ideal power transmission. If region r is the importing region, the power transmitted from region r' to region r should be positive and equal to ideal power transmission minus transmission loss.

$$\text{tcrr}_{t,r,r'} = \begin{cases} \text{ideal tcrr}_{t,r,r'} & (r \text{ is the exporting region}) \\ \left[1 - \text{TRL}_{r,r'}\right] \times \text{idealtcrr}_{t,r,r'} & (r \text{ is the importing region}) \end{cases} \tag{7.15}$$

The ideal power transmission from region r' to region r is equal to the negative of the ideal power transmission from region r to region r', so adding the two values would yield zero.

$$\text{ideal tcrr}_{t,r,r'} = -\text{ideal tcrr}_{t,r',r} \tag{7.16}$$

7.2.3.2 Installed Capacity

First-category power plants (NGCC, NU, HD, WD, BM, and PV) are decommissioned at the end of their expected lifetimes. Consequently, the installed capacity of type g power plants in region r in year t can be calculated as follows:

$$ic_{g,t,r} = \sum_{t'=t-T_g+1}^{t} nb_{g,t',r} \tag{7.17}$$

Power plants belonging to the second category (PC and IGCC) may be retrofitted with CCS or decommissioned before reaching the end of their expected lifetimes. Here, the installed capacity of type g power plants in region r in year t can be calculated as follows:

$$ic_{g,t,r} = \sum_{t'=t-T_g+1}^{t} nb_{g,t',r} - \sum_{t'=t-T_g+1}^{t} \sum_{t''=t'+1}^{t} \left(rf_{g,t',r}^{t''} + er_{g,t',r}^{t''} \right) \tag{7.18}$$

$$nb_{g,t',r} = er_{g,t',r}^{t'+1} + rf_{g,t',r}^{t'+1} + rm_{g,t',r}^{t'+1} \tag{7.19}$$

$$rm_{g,t',r}^{t''} = er_{g,t',r}^{t'+1} + rf_{g,t',r}^{t''+1} + rm_{g,t',r}^{t''+1} \tag{7.20}$$

The third-category power plants (PCC and IGCCC) can be constructed from new or created from retrofitting. Thus, the installed capacity of type g power plants in region r in year t can be calculated as follows:

$$ic_{g,t,r} = \sum_{t'=t-T_g+1}^{t} nb_{g,t',r} + \sum_{t'=t-T_g+1}^{t} \sum_{t''=t'+1}^{t} \left(rf_{g',t',r}^{t''} \cdot \lambda_{g',t'}^{g} \right) \tag{7.21}$$

An upper limit of total installed capacity for each type of power technology in each region is considered as follows:

$$ic_{g,t,r} \leq IC_{g,r}^{ub} \tag{7.22}$$

For the available annual new-build capacity for each technology, there exists an upper limit that is subject to labour, manufacturing capacity, and other factors. This upper limit can be expressed as follows:

$$nb_{g,t,r} \leq NB_{g,r}^{ub} \tag{7.23}$$

7.2.3.3   Fuel Supply

The annual fuel supply (relating to coal, natural gas and uranium) is subject to factors such as ultimate resources and production ability and should not exceed the upper limit of capability.

$$\sum_r tfd_{f,t,r} \leq FSC_f^{ub} \tag{7.24}$$

Chapter 8
China's Future Power Infrastructure

Demand for electricity in China has been accelerating in recent years and in a manner that has led to some significant mismatches between the distribution of power demand, generation and primary energy resources across China. Understanding how to optimally overcome this issue requires a holistic and integrated approach to the strategic planning and development of China's power sector. Material benefits could be realised by ensuring that the long-term development of the power system is optimised by considering the different regional dynamics and characteristics. In this chapter, we showcase how a multi-region optimisation model can deliver insights into the long-term optimal development pathway for China's power sector that minimises total system costs, yet reflects the regional variations in resource availability, inter-region power transmission capacity and demand. A multi-region approach arguably better reflects and considers real world conditions and challenges. In comparing results between single- and multi-region optimisations, it is possible to demonstrate how investment decisions would differ once regional differences are taken into account. In addition, we also introduce seasonal and diurnal (time of day) modelling which better reflects reality and sees different results emerge which has implications for system planning decisions.

8.1 Introduction

Demand for electricity in China has accelerated in recent years with national power generation increasing from 1347 TWh in 2000 to 5420 TWh in 2013 (NBSC 2014) with an average annual growth rate of approximately 12 %. Under a low-carbon scenario, electricity generation might be expected to grow to around 11,263 TWh by 2050 (Hu et al. 2009). With CO_2 emissions increasingly being recognised as a major cause of climate change, the case for making global

© Springer Science+Business Media Singapore 2016
Z. Li et al., *Informing Choices for Meeting China's Energy Challenges*,
DOI 10.1007/978-981-10-2353-8_8

reductions is growing stronger. China is currently the world's largest emitter of CO_2 emission (EIA 2015). With the power sector being its largest single contributor at around 40 % (NBSC 2014), it will inevitably become a major focus area for helping to mitigate climate change through future investments in low and zero-carbon emitting generation and carbon abatement technologies.

An approach that sought to optimise investments in reducing carbon emissions by considering the country as a single entity would have limitations by omitting the regional nuances and constraints. For instance, regional differentials in the cost of fossil fuels for power generation results in different regional carbon abatement costs. Important elements such as regional variations in electricity demand, resource availability, cost and quality, existing power generation asset fleets, and inter-region transmission links need to be taken into account. By combining all of these variables and constraints within a single, multi-region optimisation, it has been possible to deliver long-term (to 2050) insights into investment planning on a nationwide basis that would simultaneously deliver material reductions in carbon emissions to meet targets whilst optimising regional costs and resource dynamics via transmission links to deliver a least cost solution to the total national system.

This chapter proposes a multi-region superstructure optimisation model capable of optimising the development of China's power sector under range of carbon mitigation policy scenarios. The analysis informs the potential energy redistribution planning required through power transmission and fuel transportation, as well as a better spatial utilisation of renewable resources.

8.2   Modelling Process for China's Power Infrastructure

The main modelling methodology has previously been covered in Chap. 7. This chapter focuses on the data and output components.

8.2.1   Data

The baseline view of the power sector is derived from details of today's existing individual power plants, current inter-region transmission lines, a view of current and projected future costs and performance of technologies, forecast power demand growth for each region, and a carbon mitigation policy. The key parameters involved in this model include: power demand in regions, transmission losses and use of system charges, annual build limits for some technologies, operating hours, expected lifetime of power plants, efficiencies for fuel consumption rates, regional delivered fuel prices, capital cost for construction, capital cost for retrofit, variable operating costs, CO_2 emission factors, CO_2 caps and prices for regions, etc. (see Appendix B for more detail).

8.2.2 Model Solving

The linear programming solver, General Algebraic Modelling System (GAMS, GAMS Development Corporation, Washington DC, USA), is used to optimise for the range of variables and constraints.

8.3 Results and Insights into the Development of China's Power Sector

8.3.1 Power Generation Costs and Mix Outlook

The total cost of generating electricity is a function of capital costs for construction of all plant types, retrofit of pulverised coal and integrated combined-cycle (PC and IGCC) plant, operating-and-maintenance (O and M) costs, fuel costs and conversion efficiencies, and plant lifetimes. The costs of electricity generated from the ten types of plants modelled are shown in Fig. 8.1.

The outlook for China's power generation mix is shown in Fig. 8.2.

8.3.2 NGCC, NU, HD, WD, BM, and PV

Although CO_2 emission costs for natural gas combined-cycle gas turbines (NGCC) are lower than those for coal-fired technologies, its construction and fuel

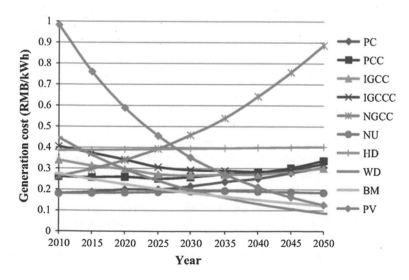

Fig. 8.1 Generation costs of different technologies

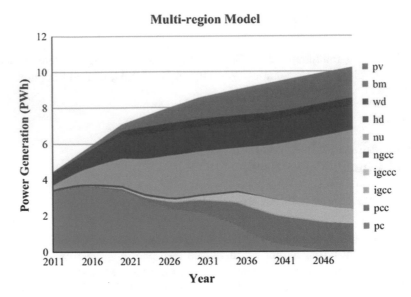

Fig. 8.2 Power generation of China's power sector. *Note* The consequence of different technologies in the figure is the same as the consequences of items of the legend

costs are much higher. Consequently, gas-fired power capacity accounts for no more than 3 % of total installed capacity today; a share that declines each year due to the high price of natural gas. Although nuclear power generation has a high capital cost, it is competitive due to long asset lifetimes, relatively low fuel costs, and zero CO_2 emissions. Consequently, annual new build rates for nuclear reach the 18 GW allowed maximum annual limit set across the planning horizon. An exception occurs during a period of anticipated supply constraint of uranium (predicted to occur between 2025 and 2035). The share of nuclear grows from 5 % in 2011 to 43 % in 2050, making it the main source of electricity generation by the end of the planning horizon. However, other factors such as safety concerns, public acceptance, and long-term waste accumulation might constrain its actual growth.

Under a cap-and-trade policy, renewable power generation technologies, e.g. hydropower (HD), wind (WD), biomass (BM), and solar photovoltaic (PV), are effective options for reducing carbon emissions and most can be deployed at scale rapidly. Hydro generation is seen to gradually rise at first and then remains almost unchanged as the resource potential becomes fully exploited. Wind and biomass power demonstrate rapid development until reaching the upper bound of resource availability in 2025. Photovoltaic power generation starts to grow rapidly in 2020 and reaches its limit of annual new build capacity in 2040. In reality, renewable technologies such as PV and WD may not develop quite as rapidly if the intermittent nature of their output becomes too de-stabilising for the national electricity grid system to cope with.

8.3.3 PC, PCC, IGCC, and IGCCC

The total generation from all coal-fired technologies, PC, pulverised coal with carbon capture and storage (PCC), IGCC, and IGCC with carbon capture and storage (IGCCC) decreases by 30 % from 3.34 PWh in 2011 to 2.36 PWh in 2050, primarily due to rising fuel prices and competition from renewable energy. The regional power generation mix of five typical regions is presented in Fig. 8.3. PC generation in each region initially sees a gradual decrease, followed by a sharp decline with the decline starting at different times among regions. It follows that, at an overall national level, total output from PC decreases gradually (Fig. 8.2). The decline starts first in those regions where the delivered price of coal is the highest, i.e. Fujian followed by South, then Shandong, East, Northeast, North, Northwest, Central, Chuanyu, and Xinjiang. This insight illustrates one advantage

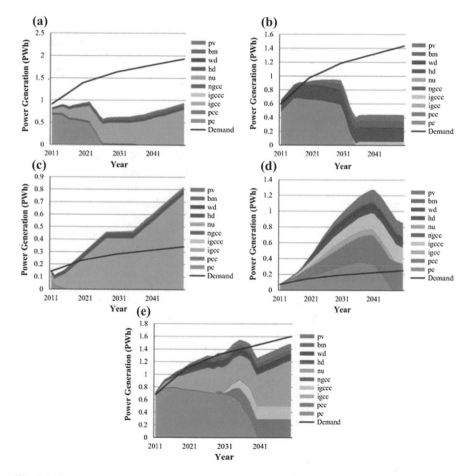

Fig. 8.3 Power generation of power sectors of typical regions. **a** East. **b** Central. **c** Fujian. **d** Xinjiang. **e** North. *Note* The *red curve* stands for regional power demand

of a multi-region model over a single, national level model. The results highlight the importance of regional variations in the cost of fuel in determining inflection points between competing technologies and meeting competition to supply fuels via inter-region transmission connections as routes to market.

Whilst the contribution of PC declines, other coal-fired technologies with CCS (PCC, IGCCC) become major contributors by 2050. Again, with regional variations in fuel costs reflected in the multi-region model, PCC and IGCCC are deployed earlier in regions facing higher coal prices; their higher efficiency and lower CO_2 emissions make them more compelling than unabated coal. In lower coal price regions, their competitive advantage takes longer to become apparent.

The share of IGCCC is less than that of PCC in the model primarily due to their different generation costs. Before 2040, the generation cost of PCC is lower than that of IGCCC, while after 2040 it is the opposite as IGCC technology matures. With supercritical and ultra-supercritical technologies more mature than IGCC today, combining it with CCS in the short to medium term is likely to be more cost-effective solution for helping to decarbonise the power sector. IGCC combined with CCS will become increasingly deployed as the technology matures in the longer term.

8.3.4 Sensitivity Analysis

The evolution of the power sector could be greatly impacted by key factors, for instance, availability of natural gas, carbon policies, etc., could, either individually or in unison, create significantly different outcomes. In this section, results are shown from impact assessments on two key variables: natural gas prices and carbon prices.

8.3.4.1 Natural Gas Price

High uncertainty exists on the future availability of gas and the prices that end-users will have to pay. Sensitivity analysis was conducted to test how different annual growth rates for natural gas prices might affect the outlook for NGCC plants. The results (see Fig. 8.4) show that NGCC's share of the generation mix is highly sensitive to different annual growth rate, particularly at 2 % p.a. and above. At 4 % p.a. and above, NGCC would virtually disappear from the generation mix altogether around 2040.

8.3.4.2 Carbon Price

Carbon price is a great uncertainty, and the sensitivity analysis on carbon policies is valuable in showing the impact that it could have on the power sector in

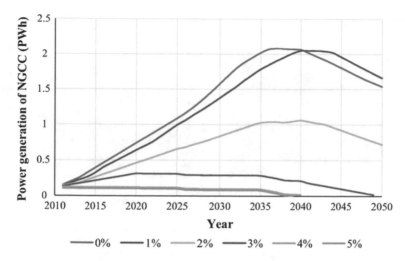

Fig. 8.4 Impact of different annual growth rates for natural gas prices on NGCC generation

changing the landscape for competing fuels and generating technologies. A baseline assumption that sees carbon pricing reaching 1000 RMB/t in 2050 was used to test the impact of the government's ambition to mitigate carbon emissions. Two further carbon price sensitivities with different annual growth rates, 5 and 0 %, were also run and the results for all three cases shown in Fig. 8.5.

The results indicate that the carbon prices can have a significant impact on changing the penetration of technologies with CCS and the timescales within which they may be required. At the higher annual growth rate for carbon prices, PCC and IGCCC both take a bigger proportion, while PC and IGCC power plants become less competitive.

8.3.5 Regional Power Generation Mix and Inter-region Power Transmission

The results of the regional power generation mix for a number of typical regions, as well as the optimal inter-region transmission pathways, are provided in this section.

The regions are classified into three types: importing regions with limited local natural resources but high demand for electricity (with positive net amount of transmission), exporting regions with abundant resources and relatively small power demand (with negative net amount of transmission), and self-sufficient regions, which have a balance between supply and demand (with net transmission around zero).

Importing regions include East, Central, and South. The power generation and demand for two typical importing regions, East and Central, are shown in Figs. 8.6 and 8.7. East China relies on imported power from regions such as Central and

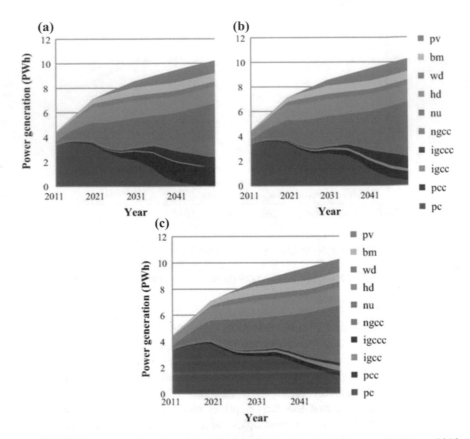

Fig. 8.5 National generation mix under different carbon price growth rate sensitivities. **a** 2016: 50 RMB/tCO$_2$; 2016–2020: 5 %; 2020–2050: 10 %. **b** 2016: 50 RMB/tCO$_2$; 2016–2050: 5 %. **c** 2016: 50 RMB/tCO$_2$; 2016–2050: 0 %

Fig. 8.6 Net amount of power transmission of typical regions

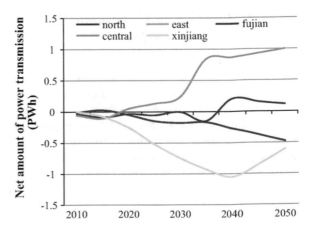

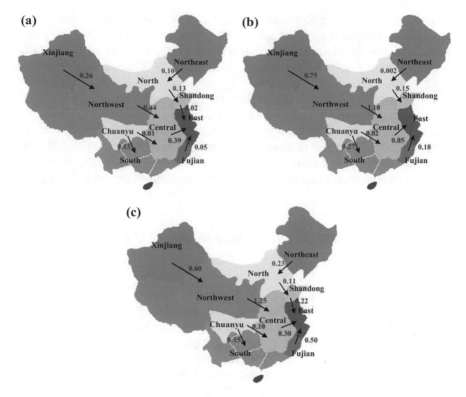

Fig. 8.7   Power transmission pathway (PWh). **a** 2020, **b** 2030 and **c** 2050

Fujian for three main reasons: high imported coal prices, large power demand, and lack of resources. With coal and hydropower resources more abundant in Central China than East China, the building of nuclear power in these regions is prohibited, being restricted to coastal regions only. Thus, Central China is dependent on imported power from Northwest and Chuanyu. The proportion of net imported power to power demand reaches 70 % by 2050 and the generation by 2050 only at 67 % of the generation in 2011.

Exporting regions include Xinjiang, Northwest, Fujian, and Chuanyu, some of which begin to export over half of their generated power to other regions by the end of the planning horizon. The power generation and the demand of two typical exporting regions, namely, Fujian and Xinjiang, are shown in Figs. 8.6 and 8.7. For Fujian Province, the cost for technologies other than nuclear power is high due to the lack of fossil fuels and renewable resources. Consequently, nuclear power generation makes up over 90 % throughout the planning horizon. Low electricity demand makes Fujian a main exporting region towards East China. Xinjiang, with its small population and vast reserves of fossil fuels and renewable resources, exports 83 % of its power generation to Northwest, then on to other regions. The dominant technology in Xinjiang is coal-fired technology whose generation is almost 80 % of the total due to very low cost coal. After 2040, coastal areas become less dependent on importing power because of the rise of nuclear power that occurs after the period of constrained uranium supply; allowing them to start to redress supply–demand imbalances. Generation in Xinjiang decreases as a result of this.

The net amount of transmission of the five regions is shown in Fig. 8.6.

Self-sufficient regions mainly include North, Northeast, and Shandong. For North China, the generation supply and demand are virtually balanced as shown in Fig. 8.6. From the maps of transmission paths and capacity in 2020, 2030, and 2050, four main power transmission pathways of China are: Xinjiang-Northwest-Central-East, Chuanyu-South, Northeast-North-Shandong-East, and Fujian-East.

8.4 Modelling Temporal Variations of Power Demand and Generation

Having created a regional power model, the next enhancement to the model was to better reflect the realities of a power system with changing demand profiles within a year by taking into account the seasonal and diurnal changes in the demand for electricity (temporal variations) as well as the generation availability of intermittent renewable like wind and solar PV. Our analysis included a case study to compare the effect of a cap-and-trade carbon mitigation scheme when the multi-region optimisation included and excluded the factors. The results indicate that NGCC plants, which have better peak shaving (regulation) capability, would be more economical than coal-fired power plants to meet the temporal peak demand phenomenon. Nevertheless, coal-fired plants would still play an important role in peak management and their peak regulation capability could significantly affect their future development. In addition, the increased deployment of renewable energy would greatly increase the demand for flexible, dispatchable thermal power plants. It is obvious that a power system with an adequate and flexible peak regulation capability would further enable the growth of renewable energy, as intermittency could be better managed.

8.4.1 Introduction

All previous studies of China's power sector have ignored the temporal variations in electricity demand and availability of different generation types at different time periods across a year as well as within a day. Peak regulation (the term Peak Shaving is also commonly used) and managing demand variations are critical for maintaining the stability and integrity of power systems. This has traditionally been managed by the build and operation of adequate generation reserve margins—often defined as having total installed generating capacity at 10–20 %

about the peak annual demand period. Increasing deployment of intermittent renewable technologies such as wind and solar photovoltaic that can fluctuate in output severely has raised the requirement for grid flexibility. It has become increasingly necessary to plan and develop any power system with due consideration to likely future patterns of load variation and intermittent generation and how these might coincide and interact across a year and within a day. Historically, the requirement has been to build and maintain an adequate reserve margin. Looking forward, the quality of that reserve margin, particularly with regards to flexibility, will be increasingly important.

Our multi-region model was further enhanced to allow for optimised load dispatch in order to deliver meaningful insights into the development of China's power sector in the long term (from 2011 to 2050) in the context of demand variations and intermittent generating technologies. The updated model now optimises China's power sector by minimising its total costs whilst reflecting the relationship of power demand, generation and transmission both spatially and temporally.

For the purpose of reflecting temporal variations of power demand as well as availability of intermittent renewable generation over a year and across multiple years, the modelling requirements become significantly more complex. Ideally, the modelling would divide each year into 8760 h and allow for the sequential dispatch of each generating unit to meet demand. However, this would lead to significant processing requirements and lengthening optimisation solving time. Equally valid results and insights can still be achieved through a simplified representation. By analysing historical electricity demand data and availability for renewable energy both across a year and within a day, we identified that demand variations and renewable availability differences were significant during different Seasons in a given year. Variations of diurnal power demand and availability for technologies within a season, as well as hourly demand and availability within a day were much less notable. Thus, in order to reflect these variations between Seasons in a year as well as different within diurnal demand periods, two representative temporal dimensions were created: four seasons in a year and three time periods in a day. Each year was effectively divided into four seasons: Spring (from March to May), Summer (from June to August), Autumn (from September to November), Winter (from December to February). Annual electricity demand was then divided into these four seasons, and each day was then divided into three periods: low load (0:00–8:00), medium load (8:00–16:00) and high load (16:00–24:00). Altogether, twelve time blocks in a year were taken to represent different power demand and renewable availability.

With respect to the twelve-block temporal division, some technologies were allocated with different availability across seasons and within the diurnal periods. According to these characteristics, the ten types of technologies were divided into three groups. Group A included technologies whose availability was irrelevant to the time blocks, such as PC, IGCC, PCC, IGCCC, NGCC, and NU. These technologies are considered dependable and are not affected by temporal conditions; thus generally do not create intermittent and uncertainty issues. Group B included HD and BM, whose availability may be variable at a Seasonal level. Group C included

WD and PV. Dispatch of these technologies is inherently intermittent and can lead to significant Seasonal and Diurnal uncertainty. By dividing a year into 12 time blocks reflecting power demand variations and differing generation availabilities, the model can now more precisely reflect load dispatch and technology features both spatially and temporally, and give insights into optimal solutions for managing these effects.

In this work, a long-term cap-and-trade policy for carbon emission reduction was implemented to stimulate the power sector to shift to a more low-carbon system. Limits on carbon emissions were allocated to regions considering future reduction goals and historical emissions. A price for carbon was set in order to penalise regions whose emissions would exceed the limit, or benefit regions whose emissions would be less than the cap. The carbon caps and prices were set to vary over years to reflect increasingly strict carbon emission policy (for details see Appendix B).

8.4.2 Objective Equations and Constrains

Objective functions, physical constrains and other conditions of the model are presented in this section. Definitions of parameters and variables are listed in Appendix A. Six sets; including t, r, g, f, s and l stand for year, region, type of power plant, fuel type, season, and diurnal load category, respectively. A specific season and daily load category is expressed as a time block (s, l). Variables t' and t'' share the same set as t, while r' shares the same set as r.

8.4.2.1 Objective Functions

As with the previous version of the model, objective functions of this model are defined to minimise the accumulated total costs for power generating sector and power transmission from 2011 to 2050. The accumulated total costs is the sum of each year's costs, which consist of six parts, including capital cost for construction, capital cost for retrofit, operation-and-maintenance cost, fuel cost, CO_2 emission cost, and transmission cost. While calculating fuel costs and transmission costs, power generated in region r and transmitted from region r' to r in year t is the sum of its corresponding values in the 12 time blocks, which can be expressed as follows:

$$pg_{r,g,t} = \sum_{g} pg_{r,g,t}^{s,l} \tag{8.1}$$

$$tcrr_{t,r,r'} = \sum_{r'} tcrr_{t,r,r'}^{s,l} \tag{8.2}$$

8.4.2.2 Physical Constrains

Power Demand and Transmission

Instead of considering regional power balances on a year-based time scale, the model now requires power demand to be satisfied in each of the twelve time blocks in a year. Meanwhile, operating hours for technologies are set as a variable that can adjust due to the constraints of natural resource as well as the need for peak regulation, i.e. dispatch can be flexed to match the variable Seasonal and Diurnal demand profiles (previously this was fixed to reflect the dispatch pattern determined by the power price tariff mechanism). Power demand in a time block is the sum of power generated within the region and transmitted from other regions. The equation for power balance in region r in time block (s, l) can be expressed as follows:

$$PD_{r,t}^{s,l} = \sum_g pg_{r,g,t}^{s,l} + tc_{t,r}^{s,l} = \sum_g ic_{r,g,t} \cdot oh_{r,g,t}^{s,l} + tc_{r,t}^{s,l} \tag{8.3}$$

Power transmitted to region r in time block (s, l) in year t equals to the sum of actual power transmitted from the other regions to region r. This can be expressed as follows:

$$tc_{t,r}^{s,l} = \sum_{r'} tcrr_{t,r,r'}^{s,l} \tag{8.4}$$

For power transmitted from region r' to r in time block (s, l) in year t, if r' is the power exporting region, the value is positive and equals to the ideal transmission power. On the other hand, if r is the power exporting region, the value is negative and equals to the difference of ideal transmission power and transmission loss.

$$tcrr_{t,r,r'}^{s,l} = \begin{cases} idealtcrr_{t,r,r'}^{s,l} & (r \text{ is the exporting region}) \\ [1 - TRL_{r,r'}] \times idealtcrr_{t,r,r'}^{s,l} & (r \text{ is the importing region}) \end{cases} \tag{8.5}$$

The sum of ideal transmission power between two regions should yield zero in a time block (s, l) in a year, and this can be expressed as follows:

$$idealtcrr_{t,r,r'}^{s,l} = -idealtcrr_{t,r',r}^{s,l} \tag{8.6}$$

Load Factors for Technologies

In the previous time-integrated model, operating hours for different kinds of power generation were defined as parameters and modelled as a statistic average for the whole country. For instance, operating hours for coal-fired power plants were fixed at 5031 h. However, actual operating hour can reach as high as 8000 h or more.

In the latest version of the model, operating hours for the ten types of technologies are not fixed parameters but is now optimised across the twelve time blocks. Instead of discussing the variation of operating hours in time blocks, this study uses load factors to reflect the utilisation of different kinds of power plants. The limits of load factors could indicate peak regulation capacity for non-renewable technologies or represent abundance of resources for renewable energy. In terms of peak regulation ability, the greater difference between the upper and lower limits of the load factor, the better its peaking regulation capacity is. Limits for load factors and its relationship between operating hours are expressed as follows:

$$MINLF^{s,l}_{r,g} \leq lf^{s,l}_{r,g,t} \leq MAXLF^{s,l}_{r,g} \tag{8.7}$$

$$oh^{s,l}_{r,g,t} = lf^{s,l}_{r,g,t} \cdot 8760/12 \tag{8.8}$$

Other Constraints

Other constraints are consistent with those outlined in Chap. 7 for the regional model.

8.4.3 Data

Data in this study included current existing power plants, costs, and features for technologies, regional power demand forecast, and an assumed carbon market. Some parameters were imported from previous work. The other key parameters were set as follows.

8.4.3.1 Power Demand

Annual regional power demand was divided into 12 time blocks representing load variations between seasons and diurnal periods based on actual data in 2001. First, regional power demand was divided into four seasons as shown in Table 8.1, and then further divided into 12 time blocks as shown in Table 8.2.

8.4.3.2 Load Factors

Upper and lower limits of load factors for the three groups of power generation technologies were set as follows:

Table 8.1 Percentages for regional power demand in four seasons

	Spring	Summer	Autumn	Winter
Northeast	25.4	25.7	24.8	24.1
North	25.4	25.8	24.7	24.1
Shandong	25.4	25.7	24.8	24.1
East	25.5	26	24.7	23.9
Fujian	25.4	25.7	24.8	24.1
South	25.3	25.6	24.8	24.3
Chuanyu	25.6	26.4	24.5	23.5
Central	26	27.5	24.1	22.4
Northwest	25	24.5	25.2	25.4
Xinjiang	25.4	25.7	24.8	24.2

Table 8.2 Percentages for regional power demand in 12 time blocks

	Spr_l	Spr_m	Spr_h	Sum_l	Sum_m	Sum_h	Aut_l	Aut_m	Aut_h	Win_l	Win_m	Win_h
Northeast	7.0	9.0	9.3	7.1	9.3	9.4	6.9	8.7	9.2	6.7	8.4	9.0
North	7.1	9.1	9.2	7.1	9.4	9.3	7.0	8.7	9.0	6.9	8.4	8.8
Shandong	7.0	9.0	9.3	7.1	9.3	9.4	6.9	8.7	9.2	6.7	8.4	9.0
East	7.2	9.3	9.0	7.3	9.6	9.1	6.9	9.0	8.8	6.7	8.6	8.6
Fujian	7.0	9.0	9.3	7.1	9.3	9.4	6.9	8.7	9.2	6.7	8.4	9.0
South	6.6	9.2	9.5	6.7	9.3	9.6	6.5	9.0	9.4	6.3	8.7	9.2w
Chuanyu	7.0	9.1	9.5	7.2	9.4	9.8	6.7	8.7	9.1	6.4	8.3	8.8
Central	7.1	8.9	9.9	7.6	9.8	10.1	6.5	8.0	9.6	6.0	7.1	9.3
Northwest	7.6	8.6	8.7	7.5	8.5	8.5	7.7	8.6	8.9	7.7	8.6	9.0
Xinjiang	7.0	9.0	9.3	7.1	9.3	9.4	6.9	8.7	9.2	6.7	8.4	9.0

Group A (PC, IGCC, PCC, IGCCC, NGCC, and NU)

The limits of load factors for power plants in this group are irrelevant to the time blocks. As presented in Table 8.3, these plants were assumed to operate up to a maximum of 8000 h per year (allowing for planned maintenance and unforced outages), which was equivalent to a load factor of 91.3 %. In addition, it assumed that NGCC would most likely become the main approach for meeting peak demand due to its high ramp-up and ramp-down rates (proven across the world) and lower fixed costs. Consequently, its lower operational limit was set to zero reflecting its ability to cycle or two-shift. Coal-fired plants were considered to perform peak regulation as well, but the lower limit of its load factor was set at 36.7 % due to its relatively low ramp-up and ramp-down rates (the lowest utilisation rate for coal plants was set to 40 % and by excluding one month for maintenance in a year, the lower limit gave 36.7 %). Nuclear power plants were considered unsuitable for peaking duties and, therefore, their lower limit was set at 79.9 %, equivalent to 7000 operational hours per year.

Group B (HD and BM)

For this group of technologies, load factors were relevant to the Season but not the Diurnal periods. The upper limits of seasonal load factors for regional hydroelectric power were assumed to be 1.1 times the actual average regional load factors from 2004 to 2009. With respect to biomass, the upper limits in Spring, Summer, and Winter were set to be 1.1 times its actual average load factor (38.5 %) in 2008, and the Autumn figures were set to be 1.5 times the average. The upper limits of load factors for hydropower and biomass were listed in Tables 8.4 and 8.5, respectively, and the lower limits of these were assumed to be zero.

Table 8.3 Limits of load factors for Group A technologies

	PC	PCC	IGCC	IGCCC	NGCC	NU
$MAXLF_{r,g}^{s,l}$	0.913	0.913	0.913	0.913	0.913	0.913
$MINLF_{r,g}^{s,l}$	0.367	0.367	0.367	0.367	0.000	0.799

Table 8.4 Upper limits of load factors for hydropower in regions

	Spring	Summer	Autumn	Winter
Northeast	0.13	0.21	0.15	0.10
North	0.12	0.08	0.12	0.20
Shandong	0.12	0.08	0.12	0.20
East	0.24	0.27	0.19	0.07
Fujian	0.40	0.63	0.53	0.34
South	0.48	0.69	0.59	0.26
Chuanyu	0.32	0.57	0.50	0.23
Central	0.53	0.67	0.45	0.25
Northwest	0.31	0.41	0.37	0.18
Xinjiang	0.29	0.48	0.45	0.25

Table 8.5 Upper limits of load factors for biomass in the whole country

Seasons	Spring	Summer	Autumn	Winter
Load factors	0.42	0.42	0.58	0.42

Group C (WD and PV)

The load factors for this group were constrained by the Seasons and Diurnal periods owing to their inherent intermittency and uncertainty. Based on the load factor for wind (23.9 %) in 2010, certain ratios are multiplied to determine its upper limits (as shown in Table 8.6). These ratios demonstrated characteristics of wind resources, such as strong wind in Spring and Winter as well as during the overnight periods. In terms of PV, the base load factor was 19.4 % with certain ratios multiplied to determine its upper limits (as shown in Table 8.7). Maximum Seasonal operating hours is the sum of corresponding maximum values in the three diurnal periods. For instance, maximum operating hours for type g technology in Spring is the sum of maximum operating hours in spr_l, spr_m, and spr_h. The lower limits for this group were set to be zero.

8.4.4 Results and Discussions

The work mainly aimed for revealing the effects of temporal division on future installed capacity trends and availability for different technologies. The role of coal-fired and NGCC plants in peaking regulation were also discussed.

8.4.4.1 Installed Capacity and Operating Hours for Technologies

In order to examine the influence of temporal division on power planning, this part not only gave the results of installed capacity and operating hours in the load

Table 8.6 Multiplier ratios for determining the upper limit of load factors for WD

Time blocks	Spr_l	Spr_m	Spr_h	Sum_l	Sum_m	Sum_h
Ratios	1.8	0.9	1.8	0.84	0.42	0.84
Time blocks	Aut_l	Aut_m	Aut_h	Win_l	Win_m	Win_h
Ratios	0.84	0.42	0.84	1.8	0.9	1.8

Table 8.7 Multiplier ratios for determining the upper limit of load factors for PV

Time blocks	Spr_l	Spr_m	Spr_h	Sum_l	Sum_m	Sum_h
Ratios	0	3.3	0	0	4.5	0
Time blocks	Aut_l	Aut_m	Aut_h	Win_l	Win_m	Win_h
Ratios	0	3.3	0	0	2.1	0

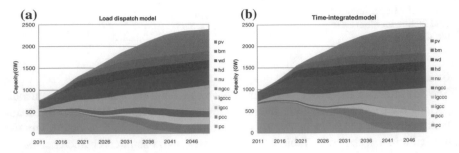

Fig. 8.8 Installed capacity of China by 2050 using the two models. **a** Load dispatch model. **b** Time-integrated model

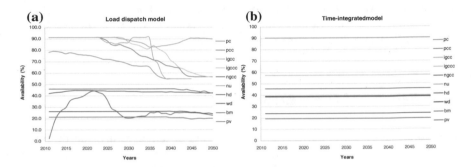

Fig. 8.9 Utilisation for technologies using the two models. **a** Load dispatch model. **b** Time-integrated model

dispatch model, but also compared to those in the time-integrated model. Installed capacity in the whole country for the two models was shown in Fig. 8.8a, b, respectively. From these figures, it can be seen that the capacity of coal-fired power plants and NGCC would vary a lot in the two models. By contrast, capacity trends for NU, HD, WD, BM and PV would share the same patterns. In addition, national annual availability for technologies (equals to average annual operating hours among regions divided by 8760 h in a year) in the two models were illustrated in Fig. 8.9a and b, respectively. Availability for thermal plants in the two models showed significant differences: flat in the time-integrated model, but fluctuated in the other one. On the other hand, availability for the other technologies showed similar values in the two models. Analyses for the results were further discussed in the next part.

8.4.4.2   Impact on Coal-Fired Plants: PC

Installed capacity of coal-fired plants varied significantly between the two models once we had incorporated the detailed temporal division of power demand and

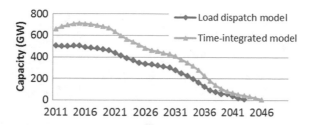

Fig. 8.10   Installed capacity for PC plants in the two models

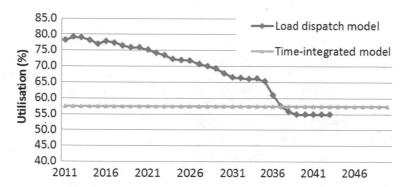

Fig. 8.11   Utilisation for PC plants in the two models

output. Installed capacity and utilisation for PC plants in the two models were illustrated in Figs. 8.10 and 8.11, respectively.

PC capacity in the load dispatch model would shrink significantly compared to that in previous model. The reason for this is that the load dispatch model is able to determine the optimal dispatch of PC plant within its lower and upper utilisation boundaries. In the time-integrated model, utilisation was fixed at a historic average of 57.4 %. This insight demonstrates how a centrally planned pricing mechanism can quickly lead to inefficient investment and unnecessary overcapacity. China operates a fixed tariff pricing system for generators that is set centrally and designed to deliver a price that enables plants to recover costs over their lifetimes. This effectively creates an incentive for China's PC fleet to deliver, on average, an actual utilisation profile that matches the utilisation level assumed in the calculation of the power price tariff (around 60 %). By fixing plant utilisation at historic averages, the time-integrated model effectively reflects the fixed tariff regulated market. When, under the load dispatch model, PC is allowed to generate at higher utilisation levels, less capacity needs to be built to deliver the same amount of electricity demand (Fig. 8.10). In allowing the model to determine the optimal utilisation across all plants, the load dispatch model better reflects a free market mechanism whereby supply and demand are allowed to interact to determine the new build and utilisation levels of different assets. The lower installed capacity of PC plant in the load dispatch model demonstrates one way in which China's regulated market mechanisms can contribute to its overcapacity issues.

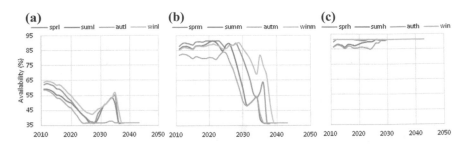

Fig. 8.12 Utilisation for PC plants in the load dispatch model

The utilisation of PC plants arising from the load dispatch model is shown in Fig. 8.12. Seasonal utilisation was high in Winter, becoming progressively lower during the Spring, Summer and Autumn. This is mainly due to a combination of lower levels of renewable availability and higher demand levels in Winter and Spring, versus relatively abundant availability of renewable generation and relatively lower demand levels in Summer and Autumn. The diurnal utilisation of some PC plants can fluctuate significantly reflecting its involvement in peak regulation. In addition, utilisation in the medium and high load periods are similar in the initial years, but start to differ significantly in the low load periods by 2025. Utilisation in the medium load period begins to drop dramatically and approach that of the low load period in the long term.

The results indicate that solar PV is likely to be deployed quickly and at scale in the future. Due to the intermittent nature of solar power, its generation is mainly limited to the medium load block and to the sunnier seasons. To ensure that peak demand can still be met, the installed capacity of PC plants is not greatly affected by the build out of solar PV. However, its utilisation declines over time as it is increasingly displaced by solar PV when it is able to generate.

8.4.4.3 Impact on the Deployment of NGCC Plants

Figure 8.13 shows the installed capacity of NGCC as determined by the two models. It reveals significant differences for future investment/deployment pathways.

Fig. 8.13 Installed capacities for NGCC plants in the two models

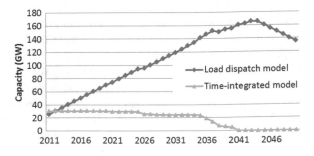

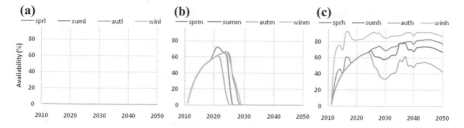

Fig. 8.14 Utilisation for NGCC plants in the load dispatch model. **a** Low load. **b** Medium load. **c** Peak load

The load dispatch model projects that NGCC capacity would expand steadily and play an important role in meeting peak demand and helping manage intermittent renewable. In contrast, NGCC plays a very limited role in the previous time-integrated model. Load factors for NGCC in the load dispatch model were allowed to vary between 0 and 91.3 % with the requirement of peak regulation introduced by the segmentation of three diurnal demand periods. NGCC offer a more flexible means of peak shaving than PC plants. Their lower fixed costs can also make them a better economic choice.

Utilisation for NGCC plants were modelled for peak regulation in the twelve time blocks (as shown in Fig. 8.14). In terms of the low-load blocks (Fig. 8.14a), NGCC plants would not generate electricity due to its high fuel costs.

For the medium load blocks (Fig. 8.14b), NGCC plants would play a major role until 2030. After that year, assuming the development of solar PV, which is set to only generate electricity in the medium load blocks and certain seasons, NGCC would increasingly be displaced.

With respect to the high load blocks (Fig. 8.14c), NGCC would remain at high utilisation levels indicating its important role in peak regulation. Especially in Winter, its utilisation would reach above 80 %. These results demonstrate the importance of NGCC in peak regulation. Should solar PV not develop as fast, then NGCC become even more important and would retain its place in helping to meet the medium load segment.

8.4.5 Summary of Impacts of Temporal Demand on Power Sector Planning

In this section, we have described how the load dispatch model can better be used to optimise the development of China's power sector from 2011 to 2050 by minimising the total costs over the period. The model considered the variation of power demand and technology availability in twelve time blocks, and performed a case study assessing the influence of temporal division on installed capacity for technologies as well as their utilisation. Several conclusions can be drawn from the results as follows.

For coal-fired power plants, the level of installed capacity is considerably lower when the model is allowed to determine and optimisation rates (the load dispatch model) than when utilisation rates are restricted to historic averages (time-integrated model). In addition, with a rapid growth of renewable energy, utilisation for coal and gas-fired plants would reduce as they are displaced by zero-carbon wind and solar PV technologies that can produce electricity at lower cost.

NGCC plants are projected to play a significant role in the load dispatch model, particularly with ever increasing amounts of intermittent renewable and the requirement to meet peak demand at minimal cost. Compared to coal-fired power plants, NGCC plants have greater flexible performances and have economics better suited to lower levels of utilisation.

Consequently for China, our modelling indicates that any design of a future power system with increasing intermittent renewable and wants to meet peak demand as economically as possible will need to incorporate NGCC. This insight only emerges when modelling of temporal demand and optimised plant utilisation is carried out. The resulting power system would have better peak regulation capacity and further support the deployment of renewable energy.

Appendix A: Nomenclature

See Tables 8.8 and 8.9.

Table 8.8 Physical meanings of parameters

Symbol	Unit	Physical meaning
$CAP_{g,t}^{nb}$	RMB/kW	Capital cost for construction of 1 kW capacity of power plants of type g in year t
$CAP_{g,t}^{rf}$	RMB/kW	Capital cost for retrofit of 1 kW capacity of power plants of type g (PC/IGCC) in year t
$CECAP_{t,r}$	tCO_2	CO_2 emission cap in region r in year t
CEI_f	tCO_2/t (tCO_2/Nm^3 for NGCC)	CO_2 emission factor of fuel f
CP_t	RMB/tCO_2	CO_2 emission price in year t
$FCR_{f,g,t}$	t/kWh (Nm^3/kWh for NGCC)	Consumption rate for fuel f by plants of type g in region r in year t for generating 1 kWh electricity
$FP_{f,t,r}$	RMB/kW (RMB/Nm^3 for NGCC)	Price of fuel f in region r in year t
FSC_f^{ub}	t	Upper limit of supply capability for fuel f
I	%	Discount rate
$IC_{g,r}^{ub}$	kW	Upper limit of total installed capacity of power plants of type g in region r
$MAXLF_{r,g}^{s,l}$	%	Maximum load factors for power plants of type g during time block l in season s, in region r

(continued)

Table 8.8 (continued)

Symbol	Unit	Physical meaning
$MAXLF_{r,g}^{s,l}$	%	Minimum load factors for power plants of type g during time block l in season s, in region r
$NB_{g,r}^{ub}$	kW	Upper limit of annual building capacity of plants of type g in region r
OH_g	h	Annual operational hours of power plants of type g
$PD_{r,t}$	kWh	Power demand in region r in year t
$PD_{r,t}^{s,l}$	kWh	Power demand during time block l in season s, in region r in year t
T_g	yr	Expected lifetime of power plants of type g
$TRC_{r,r',t}$	RMB/kWh	Loss cost of transmission from region r' to region r in year t
$TRL_{r,r',t}$	%	Loss ratio of power transmitted from region r' to region r in year t
$\lambda_{g,t'}^{g'}$	%	Capacity ratio of retrofitted plants of type g' (PCC/IGCCC) to the original plants of type g (PC/IGC)
$\mu_{g,t}$	%	Ratio of annual O and M cost by the total capital cost of plants of type g in year t

Table 8.9 Physical meanings of variables

Symbol	Unit	Physical meaning
atc	RMB	Accumulated total cost paid by the power sector over the planning horizon
$cec_{t,r}$	RMB	Total CO_2 emission cost paid (or profit gained) by the power sector in region r in year t
$er_{g,t,r}^{t'}$	kW	Early retiring capacity of power plants of type g (built in year t) in region r in year t'
$fc_{f,t,r}$	RMB	Cost for fuel f in region r in year t
$fd_{f,t,r}^{g}$	t(Nm^3)	Demand for fuel f by power plants of type g in region r in year t
$ic_{g,t,r}$	kW	Installed capacity of power plants of type g in region r in year t
$idealtcrr_{t,r,r'}$	kWh	Ideal power transmission from region r' to region r in year t (regardless of transmission loss)
$idealtcrr_{t,r,r'}^{s,l}$	kWh	Ideal power transmission from region r' to region r during time block l in season s, in year t (regardless of transmission loss)
$inv_{g,t,r}^{nb,er}$	RMB	Capital cost for construction of type g of power plants retired early in region r in year t
$inv_{g,t,r}^{nb,rf}$	RMB	Capital cost for construction of type g of retrofitted plants in region r in year t
$inv_{g,t,r}^{nb,nr}$	RMB	Capital cost for construction of type g of power plants retired at the end of expected lifetime in region r in year t
$inv_{g,t,r}^{nb}$	RMB	Capital cost in region r in year t for building power plants of type g

(continued)

Table 8.9 (continued)

Symbol	Unit	Physical meaning
$inv^{rf}_{g,t,r}$	RMB	Capital cost in region r in year t for building power plants of Type g that are retrofitted with CCS
$lf^{s,l}_{r,g}$	%	Load factors for power plants of type g during time block l in season s, in region r
$nb_{g,t,r}$	kW	Newly built capacity of power plants of type g in region r in year t
$om_{g,t,r}$	RMB	O and M cost of power plants of type g in region r in year t
$oh^{s,l}_{r,g,t}$	h	Operational hours of power plants of type g during time block l in season s, in region r in year t
$pg_{r,g,t}$	kWh	Power generation of power plants of type g in region r in year t
$pg^{s,l}_{r,g,t}$	kWh	Power generation of power plants of type g during time block l in season s, in region r in year t
$ptrc_{t,r}$	RMB	Total transmission cost paid by power sector in region r in year t
$rc_{t,r}$	RMB	Total cost paid by the power sector in region r in year t
$rf^{t'}_{g,t,r}$	kW	Retrofitted capacity of power plants of type g (built in year t) in region r in year t'
$rm^{t'}_{g,t,r}$	kW	Existing or left capacity of power plants of type g (built in year t) in region r in year t'
$tc_{t,r}$	kWh	Total power transmission into region r in year t
$tc^{s,l}_{t,r}$	kWh	Total power transmission into region r during time block l in season s in year t
$tce_{t,r}$	tCO$_2$	Total CO$_2$ emissions by power sector in region r in year t
$tcrr_{t,r,r'}$	kWh	Actual power transmission from region r' to region r in year t (concerning transmission loss)
$tcrr^{s,l}_{t,r,r'}$	kWh	Actual power transmission from region r' to region r during time block l in season s in year t (concerning transmission loss)
$tfc_{t,r}$	RMB	Total fuel cost in region r in year t
$tfd_{f,t,r}$	t(Nm^3)	Total demand for fuel f by power plants in region r in year t
$tinv^{nb}_{t,r}$	RMB	Total capital cost for construction of all types of power plants in region r in year t
$tinv^{rf}_{t,r}$	RMB	Total capital cost for retrofit of PC and IGCC in region r in year t
$tom_{t,r}$	RMB	Total O and M cost of all types of power plants in region r in year t

Appendix B: Parameters

See Tables 8.10, 8.11, 8.12, 8.13, 8.14, 8.15, 8.16, 8.17, 8.18, 8.19, 8.20, 8.21, 8.22, 8.23, 8.24, 8.25 and 8.26.

Table 8.10 Fuel consumption rates ($FCR_{f,g,t}$) of power plants at selected time points over the planning horizon

Plants type	Unit	2010	2020	2030	2050
PC	g-coal/kWh	319.00	294.00	276.50	264.00
PCC	g-coal/kWh	425.33	352.52	310.67	277.60
IGCC	g-coal/kWh	291.00	281.00	273.00	263.00
IGCCC	g-coal/kWh	359.26	317.16	292.92	269.47
NGCC	Nm^3-ng/kWh	0.1940	0.1885	0.1845	0.1785
NU	t-natural-U/year/GW	10,000	10,000	10,000	4633

Notes Fuel consumption rate of PCC/IGCCC is a function of fuel consumption rate of PC/IGCC and capacity ratio

Table 8.11 Expected technical lifetimes (Tg) of power plants of all types

Plants type	**PC**	**PCC**	**IGCC**	**IGCCC**	**NGCC**
T_g(years)	30	30	30	30	30
Plants type	**NU**	**HD**	**WD**	**BM**	**PV**
T_g(years)	60	70	20	20	20

Table 8.12 CO_2 emissions intensity of fuel (CEI_f)

Fuel type	Coal (kg-CO_2/kg-coal)	Natural gas(kg-CO_2/Nm^3-coal)	Uranium
CEI_f	2.78124	2.19362	0

Table 8.13 Capital cost for power plants construction over the planning horizon

Plants type	Unit	PC	PCC	IGCC	IGCCC	NGCC
In 2010	RMB/kW	4478	8285	12013	15755	3267
Annual increasing rate	%	−1.5	−3	−3	−3.5	−1
Plants type	**Unit**	**NU**	**HD**	**WD**	**BM**	**PV**
In 2010	RMB/kW	13662	13780	7912	7840	14134
Annual increasing rate	%	0	0.11	−4	−2	−5

Table 8.14 Operation-and-Maintenance (O and M) costs of power plants

$\mu_{g,t}$	0.03

Table 8.15 Capacity ratio of retrofitted plants of PC and IGCC plants

Plants type	2010	2020	2030	2050
PC	0.750	0.834	0.890	0.951
IGCC	0.810	0.886	0.932	0.976

Table 8.16 Capital cost for retrofit over the planning horizon

Plants type	Unit	2010	2020	2030	2050
PC	RMB/kW	2425	2186	1816	833
IGCC	RMB/kW	1570	1103	801	173

Table 8.17 Upper bound of annual fuel supply capability

Fuel type	Coal (Gtce)	Natural gas (TNm^3)	Uranium (kt)
	1.5	0.5	50

Table 8.18 Prediction of power demand in each region

	2010 (TWh)	2011–2020 Growth rate (%)	2020 (TWh)	2021–2030 Growth rate (%)	2030 (TWh)	2031–2050 Growth rate (%)	2050 (TWh)
Northeast	381	4.94	617	1.65	726	0.82	855
North	638	5.64	1103	1.88	1329	0.94	1602
Shandong	330	5.76	578	1.92	699	0.96	846
East	856	4.95	1388	1.65	1635	0.83	1927
Fujian	132	5.85	232	1.95	282	0.97	342
South	689	5.13	1137	1.71	1348	0.86	1598
Chuanyu	218	5.51	372	1.84	446	0.92	535
Central	556	5.81	977	1.94	1184	0.97	1435
Northwest	268	6.02	480	2.01	586	1.00	715
Xinjiang	66	8.19	145	2.73	190	1.36	250
Total	3543		5993		8227		10316

Table 8.19 Transmission loss ratio (%)

	Northeast	North	Shandong	East	Fujian	South	Chuanyu	Central	Northwest	Xinjiang
Northeast	–	4.1	–	–	–	–	–	–	–	–
North	4.1	–	2.0	–	–	–	–	3.6	2.7	–
Shandong	–	2.0	–	2.3	–	–	–	2.3	–	–
East	–	–	2.3	–	2.0	–	–	2.1	–	–
Fujian	–	–	–	2.0	–	3.6	–	2.2	–	–
South	–	–	–	–	3.6	–	2.9	3.3	–	–
Chuanyu	–	–	–	–	–	2.9	–	3.0	1.9	–
Central	–	3.6	2.3	2.1	2.2	3.3	3.0	–	3.5	–
Northwest	–	2.7	–	–	–	–	1.9	3.5	–	4.9
Xinjiang	–	–	–	–	–	–	–	–	4.9	–

Table 8.20 Transmission price (RMB/MWh)

	Northeast	North	Shandong	East	Fujian	South	Chuanyu	Central	Northwest	Xinjiang
Northeast	–	30	–	–	–	–	–	–	–	–
North	30	–	15	–	–	–	–	27	20	–
Shandong	–	15	–	17	–	–	–	17	–	–
East	–	–	17	–	15	–	–	16	–	–
Fujian	–	–	–	15	–	27	–	17	–	–
South	–	–	–	–	27	–	22	25	–	–
Chuanyu	–	–	–	–	–	22	–	23	14	–
Central	–	27	17	16	17	25	23	–	26	–
Northwest	–	20	–	–	–	–	14	26	–	37
Xinjiang	–	–	–	–	–	–	–	–	37	–

Table 8.21 Annual operational hours

Technology	PC	PCC	IGCC	IGCCC	NGCC
Annual operational hours (h)	5031[a]	5031[b]	5031[b]	5031[b]	4000[c]
Technology	NU	HD	WD	BM	PV
Annual operational hours (h)	7924[a]	3429[a]	2097[a]	3372[d]	1700[d]

Notes
[a]2010 China Electricity Regulatory Report released by the State Electricity Regulatory Commission
[b]Assuming PCC, IGCC and IGCCC have the same annual operational hours as PC
[c]Considering peak shaving, annual operational hours of NGCC was assumed shorter than PC
[d]Compiled Power Industry Statistics of 2008

Table 8.22 Regional coal prices in 2010

	Northeast	North	Shandong	East	Fujian	South	Chuanyu	Central	Northwest	Xinjiang
Price of coal equivalent (RMB/tce)	657	612	869	884	1058	919	572	649	561	256

Table 8.23 Regional natural gas price in 2011

	Northeast	North	Shandong	East	Fujian	South	Chuanyu	Central	Northwest	Xinjiang
Price of natural gas (RMB/m^3)	2.154	1.875	1.923	1.808	2.808	2.654	1.323	1.731	1.115	1.054

Table 8.24   Regional CO_2 cap (Gt)

Year	2016	2020	2030	2050
Northeast	0.380	0.396	0.442	0.440
North	0.517	0.538	0.602	0.598
Shandong	0.271	0.282	0.315	0.313
East	0.494	0.514	0.575	0.571
Fujian	0.076	0.079	0.089	0.088
South	0.402	0.418	0.467	0.465
Chuanyu	0.200	0.208	0.233	0.231
Central	0.450	0.468	0.523	0.520
Northwest	0.164	0.170	0.190	0.189
Xinjiang	0.064	0.067	0.075	0.075
Total	3.02	3.14	3.51	3.49

Table 8.25   Upper limit of total installed capacity of renewable energy (GW)

	Hydro[a]	Wind[b]	Solar[c]	Biomass[d]
Northeast	15.3	60.6	106.5	71.1
North	6.5	40.9	75.2	26.2
Shandong	0.1	3.9	105.2	24.8
East	7.6	6.5	33.7	22.1
Fujian	9.7	1.4	13.0	3.5
South	140.4	8.3	106.8	30.4
Chuanyu	111.5	4.4	61.3	28.4
Central	53.6	11	71.2	37.6
Northwest	32.4	39.4	125.6	13.7
Xinjiang	15.7	34.3	130.9	9.0
Total	392.8	210.7	829.4	266.8

Notes
[a]Books of China's renewable energy development strategy research-comprehensive volume
[b]The 3rd wind energy survey by China Meteorological Administration
[c]Assuming 0.1 % area of each region is used for PV, the efficiency of which is set as 15 %, and 70 % of daily sunshine can be used
[d]Energy consumption set as 500 gce/kWh

Table 8.26 Upper limit of annual building capacity (GW)

	Northeast	North	Shandong	East	Fujian	South	Chuanyu	Central	Northwest	Xinjiang	Total
PC	10	10	5	10	2.5	10	2.5	10	10	2.5	72.5
PCC	10	10	5	10	2.5	10	2.5	10	10	2.5	72.5
IGCC	5	5	2.5	5	1.5	5	1.5	5	5	1.5	37
IGCCC	5	5	2.5	5	1.5	5	1.5	5	5	1.5	37
NGCC	5	5	5	5	2.5	5	2.5	5	5	2.5	42.5
NU	3	3	3	3	3	3	0	0	0	0	18
HD	1.2	0.5	0.0	0.6	0.7	10.7	8.5	4.1	2.5	1.2	30
WD	8.6	5.8	0.6	0.9	0.2	1.2	0.6	1.6	5.6	4.9	30
BM	2.7	1.0	0.9	0.8	0.1	1.1	1.1	1.4	0.5	0.3	9.9
PV	3.9	2.7	3.8	1.2	0.5	3.9	2.0	2.6	4.6	4.8	30

References

EIA (U.S. Energy Information Administration) International energy statistics. EIA, Washington DC. www.eia.gov/countries/data.cfm. Accessed 25 Feb 2015

Hu X, Chen L, Lei H (2009) China's low carbon development pathways by 2050, scenario analysis of energy demand and carbon emissions. Science Press, Beijing, China (in Chinese)

NBSC (National Bureau of Statistics of China) (2014) China energy statistical yearbook 2014. China Statistics Press, Beijing, China (in Chinese)

Chapter 9
Impact of Water Availability on China's Energy Industry

In the previous chapter, by focusing on China's power sector, we demonstrated how complex modelling can be used to deliver insights into optimal pathways for the future development and evolution. The analysis indicated how the consideration of regional and seasonal variations in natural primary energy resources, wind and solar, can materially alter the results and implications. In this chapter, we begin to explore how constraints on another natural resource, water, may impact the future development of the China's energy sector, particularly in the coal industry, that could, in turn, place hurdles in achieving optimal results in the power sector.

In 2011, 610.7 billion m^3 of water was consumed in China, 23.9 % of which was used for industrial purposes. Over half of China's industrial water consumption is accounted for by the coal industry comprising the supply chain elements of coal mining and preparation, coal-fired power generation, and coal-to-chemicals manufacturing. In addition to power demand growth, continual industrialization and urbanisation are expected to increase primary energy consumption even further; 54 % of which is expected to be met by coal (NBSC 2009; IEA 2008). Importantly, coal and water resources are unevenly and inversely distributed in China—coal reserves are mainly located in northern China, while water resources are abundant in the south but scarce in the north. This geological challenge combined with intensity of water use, continued rising demand for coal and competing growth from other industrial sectors could see water resource constraints impacting future industrial and economic development. The following preliminary analysis covers water withdrawal and consumption in China's coal supply chain, including coal mining, preparation, transport, conversion, and final disposal. We also illustrate the interaction between the water system and the coal industry and present illustrative scenarios for water usage in China's coal supply chain in 2020 and 2030, based on potential technical and policy opportunities for water savings. Finally, further research developments are proposed to further evaluate specific technology and policy opportunities, including a techno-economic assessment to highlight greatest potential benefits at lowest cost.

© Springer Science+Business Media Singapore 2016
Z. Li et al., *Informing Choices for Meeting China's Energy Challenges*,
DOI 10.1007/978-981-10-2353-8_9

The main stages of the coal supply chain are illustrated in Fig. 9.1. The supply chain comprises mining, preparation, transport, conversion, and disposal of pollutants, emissions, and waste. In each stage, water is either withdrawn directly from natural water bodies (i.e., surface or underground water) or recycled from other stages and re-used. Concerns over water issues in the coal supply chain mainly involve impacts to water withdrawal for production and operations, water recycling, wastewater treatment, and discharge/aquifer recharge back into the local water system. This section is focused on water withdrawal and re-use within the coal industry but does not cover water quality treatment requirements to ensure safe discharge/recharge.

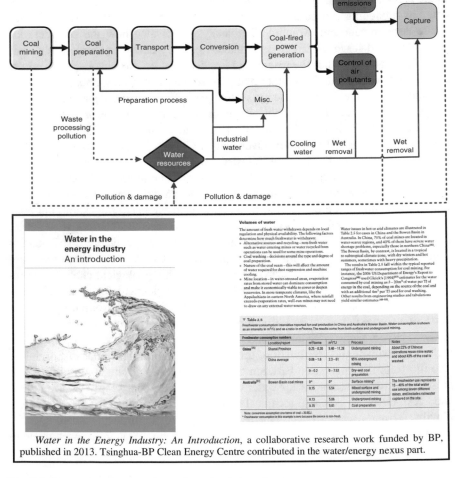

Water in the Energy Industry: An Introduction, a collaborative research work funded by BP, published in 2013. Tsinghua-BP Clean Energy Centre contributed in the water/energy nexus part.

Fig. 9.1 Interactions between the coal supply chain and water resources

9.1   Coal Industry and Water

China has 15,119 coal mines, which produced 3.63 billion tonnes of coal in 2012. Approximately 70 % of these mines are located in water-scarce areas and about 40 % experience severe water shortage problems. During the process of mining coal, 1.6–3.0 tonnes of water are consumed to produce each tonne of coal while, on average, 4 tonnes of pit water are drained. Cumulatively, coal mining produces approximately 2.2 billion tonnes of wastewater per year (NBSC 2013). Mining activities can also cause damage to water systems, especially underground water. For instance, it is estimated that around 1.07 tonnes of underground water reserves are destroyed to produce one tonne of coal in Shanxi province in China (Li et al. 2004; Wang 2010). With limited replenishment of aquifers in arid, coal-rich catchment areas, and increasing draw-down rates, depletion impacts are likely to accelerate further.

In the coal preparation sector, 94 % of China's cleaned coal is prepared via wet methods (i.e., coal washing) which plays an important role in supporting the efficiency of combustion at coal-fired power stations. Washing reduces the ash and sulphur content and helps reduce coal consumption by 2–5 gce/kWh for every 1 % decrease in ash content (Zhou 2005).

On average, for each tonne of prepared coal, 2.5 tonnes of water are used (Zhong 2001); 0.15 tonnes are removed with the by-products; 0.05 tonnes are consumed in the process and maintenance activities (Wu 2009), and the remainder is recycled in a closed loop. As of the end of 2007, there were 1300 coal preparation plants in China with a total productivity of 1.25 billion tonnes per year (Tao 2009). Still, as of 2009, only 43 % of all coal was washed, and less than 20 % of steam coal (primarily used in power generation) was washed.

Although wet preparation methods are mature and highly effective, water-related issues are one of the key barriers to increasing coal wash rates in China. The actual water loss during the preparation process is unacceptable in extremely arid regions, such as northwest China. In addition, China's abundant young coal readily degrades in water, and is thus not suitable for wet processing. After wet washing, coal's water content can rise to over 12 %, which can result in freeze-thaw disintegration when transported under severely cold conditions (Chen and Luo 2003); particularly in winter when coal demand for power generation is at its peak.

9.2   Coal Conversion into Power

Coal-fired power generation remains the dominant sector for coal conversion. Most water used in the coal industry is used during the process of converting coal, mainly in coal-fired power plants; many of which are located in arid regions. In 2005, the average water consumption rate of thermal power plants in China was

3.1 kg/kWh. Most of these were coal-fired plants, while the average water consumption of thermal power in the US at the same time was only 1.78 kg/kWh (Jiang and Han 2008).

This disparity is due mainly to the difference in the countries' respective thermal power generation fleets (EIA 2011; Mielke et al. 2010) with the US have significantly more single and combined cycle gas-fired power stations that can consume less than half that of coal-fired power stations (single-cycle gas-fired turbines require no water at all).

US coal-fired power capacity peaked at 30 % of the total generation mix (EIA 2011) and continues to fall as US gas prices remain low and renewables continue to be built and clean air legislation tightens. In China, the vast majority of thermal power plant is coal-fired.

In China's coal-fired power plants, most of the water requirements stem from the need to cool and condense steam. The ratio of solids in the water being added to the system versus the solids in the water being removed from the system has to be controlled to limit build-up of solids in cooling towers. Dry air cooling (or air-cooled condensers) can offer substantial water savings if employed (Wang 2009) but adds to capital costs and can reduce power plant efficiency by 7–8 %. However, currently only 4.25 % of coal-fired power units in China use dry air cooling, leaving an opportunity for major reductions in water usage of up to 60 %. Increased deployment of supercritical power plants and, ultra-supercritical power plants could further reduce water consumption and help offset the efficiency losses of dry air cooling. Either way, China has choices depending on its priorities. Whilst dry air cooling could significantly reduce water requirements, the reduction in efficiency would both increase demand for coal and increase CO_2/MWh emissions.

9.3 Coal Conversion into Chemicals

In the coal-to-chemicals industry, approximately 2.5 tonnes of fresh water are consumed to produce one tonne of coke. However, in the coal-to-liquids industry, an indirect coal-to-liquid plant with a capacity of 1.5 million tonnes per year consumes 10 tonnes of water to produce one tonne of product, making it an extremely water-intensive industry. Potential mitigation technologies currently available are recycling water from raw coal preparation, recycling cooling water, and treatment and recycling waste water, which could offer potential savings of 50–60 %.

In summary, the intensity of water utilisation by China's coal industry presents current and growing future challenges and while opportunities for reducing intensity are available and being deployed in China and other nations, there is substantial potential for further improvement. The following scenarios aim to provide an initial indication of the relative impacts of technology and policy interventions and highlight that water savings can be further increased through integrated planning and consideration of a supply chain/systems approach rather than industrial stages being addressed separately.

9.4   Scenario Analysis: Water Use in the Coal Supply Chain in 2020 and 2030

As outlined above, the "Water Plan", sets water use targets for China of 127 billion m^3 and 120 billion m^3 in 2020 and 2030, respectively. In its World Energy Outlook (WEO 2010), the International Energy Agency (IEA) provided three scenarios to analyze global and regional energy trends in 2020 and 2030. For this research, the coal demand data in the WEO Current Policies Scenario is taken as a base case; the data in the New Policies Scenario is used to estimate the impacts of policy on future trends. In addition, based on the present status of technology development, the effects of technology on water use are applied in the scenarios.

Building on these factors, we proposed the following four scenarios for initial analysis:

- Scenario 1 (S1) serves as a baseline, without any consideration given to policy changes and/or technology improvements.
- Scenario 2 (S2) considers the impacts of technology improvements under current policies.
- Scenario 3 (S3) considers the impacts of policy changes without technology improvements.
- Scenario 4 (S4) combines the policy changes and technical progress to account for the synergistic impacts.

The results of the scenario analysis are presented in Fig. 9.2, which illustrates water use in the coal supply chain from 2008 to 2030, as well as the estimated total industrial water use targets according to the Water Plan (where water use refers specifically to water withdrawal for power generation). In Scenario 1 and

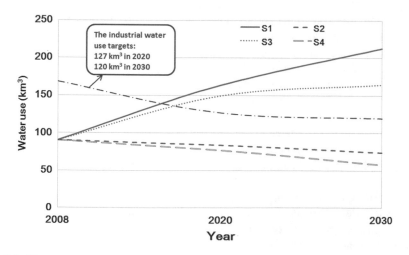

Fig. 9.2   Water use trends of the coal supply chain in China 2008–2030

Scenario 3, where technological improvements are not considered, water use in the coal supply chain continues to increase and exceeds the targets in just a few years.

(NB The 2010 State Council response to the Water Plan used 2008 as the base case year; for this reason, 2008 was also used as the base case year for this analysis).

The results from Scenario 1 indicate that if the Chinese government does not adopt any water-saving policies or promote any technological improvements, water use in the coal supply chain will continue to increase dramatically; by 2030 to double 2008 levels and exceeding the water plan as soon as 2016.

In Scenario 3, the impact of policy changes is taken into account. Water use in this scenario does not increase noticeably from 2020 to 2030. However, the projected water use under Scenario 3 still fails to achieve the targets set in the Water Plan, mainly because water-saving technologies are not deployed at sufficient scale. The results indicate that the potential capability of policy changes alone to reduce water use in the coal supply chain may be limited. To achieve necessary water savings, the government should implement even stricter policies and incentivise the development of relevant water-saving technologies.

In Scenarios 2 and 4, technology applications are introduced and water use in the coal supply chain decreases slightly before 2020 and declines appreciably after that. The results from Scenario 2 indicate that, with the current policies, developing and deploying water-saving technologies can effectively control and even reduce water use in the coal supply chain to about 66 % of the total industrial water use target. In 2020, this ratio is only slightly higher than in 2008, but between now and 2020 there is also expected growth in other water-intensive industries (e.g., paper making, metallurgy, and textile industries). Therefore, we predict that actual industrial water use will exceed the targets around 2020 if only technology improvements are applied. New effective policies are necessary to ensure the water use target is achieved. The results of Scenario 4 demonstrate the integrated influence of policies and technical progress on water use reduction. In 2030, water use in the coal supply chain would be 57.7 billion m^3, 48 % of the industrial water use target. However, around 2020, we identified the same problem with that described in Scenario 2. Under Scenario 4, water use in the coal supply chain in 2020 is projected to be 76.9 billion m^3, roughly 60 % of the industrial water use target. In this case, actual industrial water use is likely to exceed the water-saving target for the same reasons discussed above.

Comparing the results of Scenario 2 and Scenario 3, technological improvements would appear to offer the greatest potential for delivering water saving than policy changes and we propose further analysis to demonstrate how detailed technology deployment could ensure that the 2020 and 2030 industrial water use targets can be achieved. However, it is clear that a combination of policies and accelerated technology deployments may be most effective, including consideration of water pricing/water productivity measures to focus attention on which water-saving technologies/combinations of technologies would be the optimum solution for the supply chain as a whole and deliver greatest benefits for the longest time whilst minimising costs.

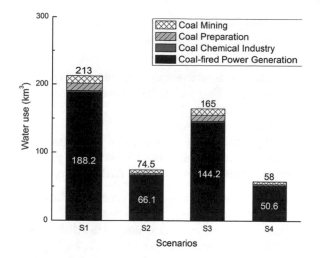

Fig. 9.3 Water use by sector in the coal supply chain in China 2030

An initial illustration of comparisons between the supply chain elements within the China coal industry under each scenario highlights is shown in Fig. 9.3. Coal-fired power generation dominates water consumption in all cases and, therefore, offers the greatest potential for water savings in absolute terms. In 2008, water use in coal-fired power generation was 78.6 billion m^3, with electricity generation of 2759 TWh. In Scenario 2, coal-fired power generation is 6605 TWh in 2030, an increase of 140 % compared to 2008, while total coal industry water use is reduced by 16 %. In Scenario 4, coal-fired power generation is 5060 TWh in 2030, an increase of 83 %, while total coal industry water use is reduced by 36 % relative to 2008 actual levels. The technology application assumptions for each scenario include increasing the concentration ratios in the cooling systems, reducing the evaporation and drift losses in cooling towers, recycling waste water, and utilising air cooling units in arid areas and seawater cooling in coastal areas. All these technologies are currently available and can be applied to newly built or retrofitted thermal power plants.

Although the absolute water savings in coal mining, coal preparation, and coal-to-chemicals industry are relatively small compared with those for coal-fired power generation, there are significant differences between the four scenarios when examining the water use reduction rate in each sector. For coal mining, water use is 11.1 billion m^3 in Scenario 1. In Scenario 3 (policy only) water use is reduced by 17.8 %. In Scenario 2, the coal mining water use is reduced by 56.5 % and in Scenario 4 reduced by 64 %, a similar profile indicating that the technology assumptions of increasing the concentration ratios in the cooling systems and reducing the evaporation and drift losses in cooling towers have the greatest potential for reducing water use which could be material within individual catchments/regions where competing demands for water exist. There may also be crossover benefits for water quality/water safety issues associated with produced water discharge/recharge.

9.5 Practical Pathways for Water Saving—Deep-Dive Research Priorities

The discussions above highlight the potential severity of water issues in the coal supply chain in China and the urgency with which these constraints require addressing; especially given that they are already limiting the development of China's coal industry and will likely exceed the Water Plan within the next 2 years. Therefore, we suggest the following pathways are subjected to further urgent deep-dive quantitative research and prioritised on the most material/least complex opportunities across the coal industry supply chain:

- Priority 1—Coal-fired power generation has the greatest potential for delivering water saving. For a circulating cooling system, the optimum concentration ratio is 4–5, whereas in China 80 % of the coal-fired units are operating with a concentration ratio lower than 3.17, highlighting ~1–2 % reduction potential. Combined with reducing the quantity of circulating/recycling water and/or utilising air cooling units in arid areas/ seawater cooling in coastal areas could also reduce fresh water consumption.
- Priority 2—In coal mining, large quantities of waste water are currently discharged. The largest potential of water savings in the coal mining stage is to recycle the mining wastewater for processes where lower water quality is acceptable—this would include increasing the ratio of washed steam coal in China to improve the efficiency of coal-fired power generation. An alternative would be to increased coal imports that have been washed at source.
- Priority 3—In the coal-to-chemicals industry, water use is likely to increase quickly as many new plants are either under construction or planned. As projects are developed, building in higher plant efficiency and wastewater treatment could contribute to water savings as well as environmental protection of the catchment areas and communities.

9.6 Suggestions for Water Challenges in Coal Industry

Whilst a range of environmental challenges continue to emerge, water may present the most material and urgent risk to the future development of China's coal industry, particularly in the vital role of power generation. The results of our analysis have indicated that, by 2030, coal-fired power generation will continue to account for the greatest fraction of water use in the coal supply chain, even when technology improvements and stricter policies are implemented. Existing technology deployment is likely to be more effective than policy alone in the near-term future for reductions in water use and further detailed analysis of deployment impacts for a range of options/combinations will help identify the most technologically feasible and economically rational pathways. However, even with implementation of all the priority suggestions included above, these are unlikely to be sufficient to meet the targets for water consumption set by the State Council.

Therefore, the question itself may actually benefit from being reframed from "how might water constrain sustainable development of China's coal industry" to "how coal-fired power generation might impact the sustainability of China's national water resources"? A wider systems approach could open up additional pathways and illustrate trade-offs for policy makers to consider, including lessons from other nations and the mistakes and successes they have made under similar constraints. This highlights the importance of deploying a systems view to highlight how combinations of technologies and policy can enable optimisation of China's finite natural resources to sustain key industries and economic growth.

References

Chen Q, Luo Z (2003) Assessment of dry methods for coal preparation (in Chinese). Coal Prep Technol 6:34–40

Energy Information Administration, US (2009) Electric power annual—existing capacity by energy source (2009). http://www.eia.gov/electricity/annual/

IEA (2010) World energy outlook 2010. OECD/IEA. http://www.worldenergyoutlook.org/media/weowebsite/2010/weo2010_es_chinese.pdf

International Energy Agency (IEA) (2008) World energy outlook 2008. OECD/IEA. www.worldenergyoutlook.org/media/weowebsite/2008-1994/weo2008.pdf

Jiang Z, Han H (2008) Water resources utilization situation and countermeasures of thermal power unit. Huadian Technol 30:1–5

Kang L, Wang J (2008) Discussion on water-saving and consumption reduction measures in thermal power plants. Electr Power Environ Prot 24:40–43

Li J, Jiang J, Wang Y (2004) Recycling on coal mining wastewater resource and circular economy. Energy Environ Prot 18(20–22):26

Mielke E, Anadon LD, Narayanamurti V (2010) Water consumption of energy resource extraction, processing, and conversion. Energy technology innovation policy discussion paper series. Harvard Kennedy School, Cambridge, MA. http://belfercenter.ksg.harvard.edu/files/ETIP-DP-2010-15-final-4.pdf

National Bureau of Statistics of China (NBSC) (2009) China statistical yearbook 2009. China Statistics Press, Beijing

NBSC (2013) National data. National Bureau of Statistics of China. Available at: http://data.stats.gov.cn/

Tao H (2009) China coal preparation technology as viewed from the energy conservation and drainage reduction. Coal Technol 28:111–112

US Energy Information Administration (EIA) (2011) Existing capacity by energy source 2009 (Table 1.2). In: Electric power annual 2009. http://pbadupws.nrc.gov/docs/ML1104/ML110410547.pdf

Wang P (2009) The 4 development stages of domestic air cooling thermal power units and the total loading capacity (in Chinese). Power Equipment 23:69–72

Wang Q (2010) Evaluation on influences of coal mining for water resources and control measures. Shanxi Hydro Technics 175:14–16

Wu S (2009) Development of Chinese coal preparation in past 30 years (in Chinese). Coal Process Compr Utilization 1–4

Zhong Y (2001) Zero draining of waste water in coal preparation industry. Ind Water Wastewater 32:43–44

Zhou C (2005) Current situation and development of slime water treatment technology in China. Gansu Sci Technol 21:142–143

Chapter 10
Future Low Carbon Options for Combustion Power Plants

Due to the ever-increasing energy demand and a strong reliance on coal, China has been working hard to identify options for diversifying its energy supply base. Meanwhile, China is also facing serious environmental problems brought about by rapidly increasing urban waste generation associated with rapid urbanisation and burgeoning consumption levels. Waste-to-energy utilisation can ease some of this environmental pressure and also help meet some energy demand. This chapter explores both the potential role that utilising energy from waste could play in meeting China's energy demand and possible technical and policy routes for unlocking that potential. An update of the current state of greenhouse policy and CCS technology status as well as demonstration projects in China are also presented in this chapter to give an insight into future hurdles and opportunities for China's efforts on greenhouse gas emission reduction.

10.1 The Energy Potential of Municipal Solid Waste

Municipal solid waste (MSW) is the solid waste generated as a result of urban daily life activities or activities to support urban daily life, as well as the solid waste specified by laws, administrative rules and regulations. The production of MSW in China has increased dramatically in recent years. In 2013, the annual collected amount of MSW in China reached 172 million tonnes (NBSC 2014), and the harmless process rate was 89 % with an average annual growth rate of 10 % in recent years.

MSW processing modes in China are currently experiencing significant transitional change. Landfill processing rates reduced from 89 % in 2002 to 68 % in 2013, while incineration increased from 3.7 to 30 % over this period. The share of compost has not been compiled in statistics separately since 2011 due to its

© Springer Science+Business Media Singapore 2016
Z. Li et al., *Informing Choices for Meeting China's Energy Challenges*,
DOI 10.1007/978-981-10-2353-8_10

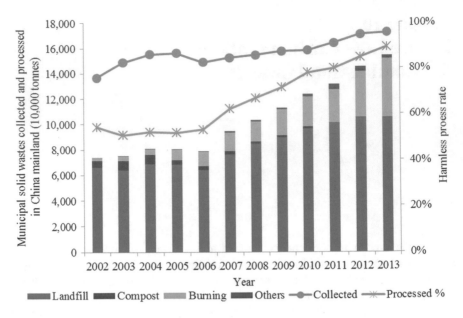

Fig. 10.1 Collected and processed MSW in Mainland China

relatively small amounts. As a predominant processing technology, landfill has a relatively low cost, but the decomposition of landfilled waste takes decades or even hundreds of years and can be detrimental to the healthy and sustainable development of land. Despite relatively extensive construction of MSW management, processing and storage facilities, many cities are facing a crisis of "garbage overload". The production of MSW not only is a threat to the environment but also shows considerable potential for energy utilisation. To date, the inherent energy potential contained in MSW remains largely untapped, as shown in Fig. 10.1.

The energy utilisation of MSW is dependent on the amount of waste production, its characteristics, and the processing and conversion technologies used. Based on the prediction of China's urban population, the economy, and the consumption level of residents, the annual MSW production rate up to 2030 has been estimated using multi-regressive methods and historical data (Yuanwei et al. 2013). The results show that the annual amount collected and transported MSW in China could increase 2–3 times in the next 20 years, reaching 3700 million tonnes in the year 2030. This is the equivalent of 63 million tonnes of standard coal (at an average calorific value of 5000 kJ/kg in China). Due to losses in collection, classification and processing, waste cannot be completely converted into useful energy. Even so, the waste to energy potential can be considered to be considerable.

10.2 Current MSW Processing Technologies

Apart from waste recycling, current MSW disposal options and technologies include sanitary landfill, incineration, compost, pyrolysis, direct gasification and plasma gasification. At present, sanitary landfill (including landfill combined with gas collection and utilisation) and incineration are the two major disposal options used in China, representing over 98 %. Although landfill dominates (about 70 % share), it suffers from problems relating to large land footprint requirements, secondary pollution, greenhouse gas emissions and slow breakdown rates. Incineration has witnessed significant growth that has benefitted from national policy support in recent years (The State Council 2012). However, compared to landfill, incineration is less economic, requiring high investment costs that make it more dependent on government subsidies to be economically viable. Landfill combined with landfill gas collection is economically challenged in the current market and requires greater policy support to be economically viable. Whilst, other MSW disposal technologies, such as compost, gasification, pyrolysis, direct gasification and plasma gasification are available, their deployment is likely to be limited by their technology maturity, availability and/or cost. With China's future MSW waste generation rates on an ever upward trajectory, the investigation on the future prospects of these technologies and their potential contribution to China's MSW sector under different policy conditions is becoming increasingly critical (Table 10.1).

10.3 Modelling of MSW Disposal Option

10.3.1 Model and Results

Based on the overview of current MSW processing technologies, an optimisation model for waste to energy was developed to study optimal pathways for China's MSW disposal sector using different potential policy instruments. The evolution of the MSW disposal sector is a dynamic process, including decisions on existing capacity to opt for continued operation, decommissioning or retrofitting and decisions on investment in new capacity with different technology choices to select from. Policy instruments such as methane emission taxes, reform of price tariffs for electricity generated from waste, waste disposal subsidies, landfill tax, etc. will directly influence these decisions and the choice of the technologies to deploy. With many potential variables and policy objectives to meet, determining an optimal pathway for China's MSW disposal sector is complex.

The objective function of the model developed is to minimise the total accumulated cost of the entire MSW system out to 2030. Four kinds of waste processing technology classes, landfill (lf), landfill with gas collection (lfgc), incineration (ic) and other technologies (oth), are taken into consideration. We assumed that all

Table 10.1 Comparison of eight waste processing technologies

Technology	Waste type requirement	Products	Processing cost (RMB/t)	Advantages	Disadvantages
Recycling	Paper, plastic, metal, glass, fabric	Roughly similar products	80	Simple processing, high economic benefit	Need support from waste sorting
Landfill	All (no need for sorting)	Methane gas	35–50	No sorting, low cost, technically reliable	Large area occupied, secondary pollution
Compost	Organic content over 40 %	Artificial fertilizer	50–80	Little secondary pollution	Low quality production, smelly
Incineration	HCV >4.186 MJ/kg, water content <50 %	Electricity, thermal heat	80–140	Large treatment capacity, reduce waste volume prominently	Secondary pollution, high processing cost
Pyrolysis	All	Fuel gas, oil	35	Productions easy to be stored or transported, low pollution	Immature in technology
Gasification	Water content 10–20 %, relatively high heat value	Fuel gas such as CH_4, H_2, CO	–	Low capital cost, high conversion rate	High facility cost
Fuelization	Combustible components after classification	Gas, liquid, solid fuel (moulded fuel)	–	Production easy to be stored or transported, high combustion efficiency, low pollution	High processing cost
Plasma	All	Syngas, thermal energy, clean alloyed slag	>200	Clean, reduce waste volume prominently	Large capital cost, high electrical energy, more CO_2 emission

facilities will be decommissioned at the end of their respective useful life times. However, the model has the option to decommission facilities early or retrofit them to another technology if it is economic to do so. The cost of construction, early decommissioning and retrofitting are considered within the total cost calculation.

The baseline view of the MSW disposal sector is derived from the installed MSW processing capacity today, a view on current and projected future costs and performance of technologies, forecast MSW disposal demand, and national policy on methane emission tax, landfill tax, waste disposal subsidies and feed-in tariffs. The optimal MSW disposal decision in the period of 2011–2030 is then obtained by solving the optimisation problem, as shown in Figs. 10.2 and 10.3. From the optimisation results, it is clear that incineration will develop steadily over next 20 years, and its share of capacity will increase to 64 % in 2030. It absorbs disposal demand formerly fulfilled by landfill, so that the annual increase rate of landfill capacity (including lf and lfgc) is slower than the actual increase rate of disposal demand.

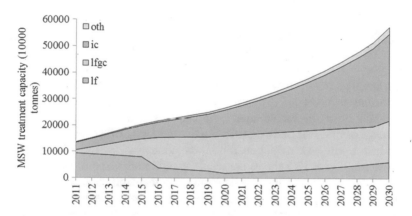

Fig. 10.2 Annual MSW total disposal capacities (base case)

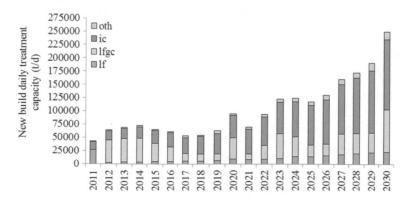

Fig. 10.3 Annual MSW daily disposal capacities (base case)

10.3.2 Sensitivity Analysis and Policy Uncertainties

In addition to creating a business as usual base case, sensitivity analysis was conducted to illustrate the effect of key factors on the central optimised result. The sensitivities included the demand for MSW disposal, waste to power efficiencies, construction costs, calorific values of MSW, landfill gas production rates, feed-in tariff, methane emissions targets and prices, and landfill taxes.

The impact of future MSW management policies on the optimal central pathway was also analysed using the model. This assessed the impact of MSW disposal subsidies, methane emission taxation and kitchen waste classification policies for better understanding the likely impact of future policy made on China's MSW disposal sector.

10.3.3 MSW Disposal Subsidy

Starting from 0.65 RMB/kWh in 2011, different linear trajectories to 2030 for the incineration feed-in tariff (resulting in prices ranging from 0.3 to 0.6 RMB/kWh by 2030) were created. A base case assumption of 0.4 RMB/kWh was also run to create a baseline. Figure 10.4 shows for the different levels of feed-in tariff and the corresponding value of the disposal subsidy required from government to ensure the profitability of the incineration companies is maintained. The results indicate a range of different outcomes from a starting point of 82 RMB/t. The annual rate of investment cost reduction for incineration was tested at levels between 0 and 4 % p.a. Figure 10.5 shows the resultant minimum disposal subsidy required across the ranges tested. The analysis shows that incineration is unlikely to maintain a long-term economic advantage when waste disposal subsidies reduce to a certain degree or investment costs fail to decrease beyond a certain point. It also demonstrates how a national feed-in tariff can be used to reduce reliance on disposal subsidies that are set at a local level and can result in significant regional disparities. This may be an important consideration that should national targets

Fig. 10.4 The minimum MSW disposal subsidy for MSW incineration plant at various electricity feed-in tariff sensitivities

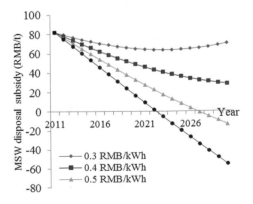

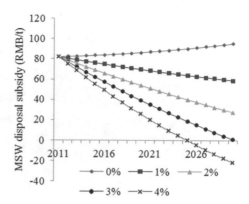

Fig. 10.5 Variations in the minimum MSW disposal subsidy required to offset different outlooks for incineration investment costs

or caps on methane emissions ever be introduced as actions at a local level could counter objectives set at national level. The model indicates that a stronger bias towards feed-in tariffs for electricity produced from MSW would allow central planners to assert more influence over investment decisions within the MSW disposal sector.

10.3.4 Methane Emission Tax

A methane emissions policy was reflected in the base case. It included a launch date, emission target and price. However, given the inherent uncertainties in projecting a policy that does not currently exist, we investigated three different approaches to policy design and compared the optimal planning results against a zero emission reduction policy scenario, as shown in Fig. 10.6. From the perspective of both reducing emissions and minimising cost, the cap-and-trade scheme is the most effective policy instrument. Even without allocating credits for over achieving against the reduction cap, allowing participants to delay investing in compliance technologies until such time that they are more mature and cheaper (and pay the penalty in the meantime) would still deliver the lowest cost solution.

10.3.5 Kitchen Waste Classification

We assumed that China launches an effective kitchen waste classification policy in 2021 and increases the portion of kitchen waste disposed and processed annually after that. Thus the contribution of other technologies is seen to reach 30 % (5 % in the base case) in 2030 with the average calorific value of waste reaching 2300 kcal/kg (2000 kcal/kg in the base case). Figure 10.7 shows the optimal

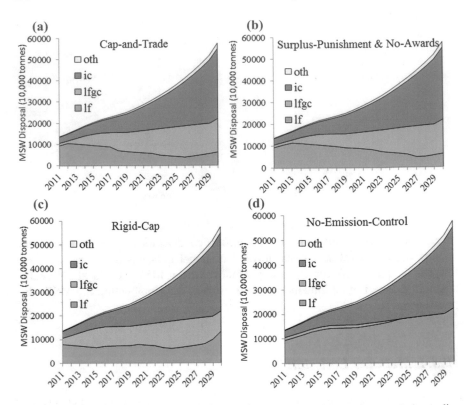

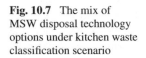

Fig. 10.6 The mix of MSW disposal technology options under four methane emission policy scenarios

Fig. 10.7 The mix of
MSW disposal technology
options under kitchen waste
classification scenario

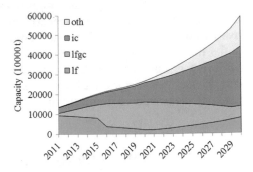

capacity of each technology under kitchen waste classification scenario. Compared with the baseline scenario, the rapid deployment of other technologies reduces the burden on landfill and landfill with gas collection, however, incineration remains the most competitive technology.

Fig. 10.8 Waste to
energy operational plant:
development by category

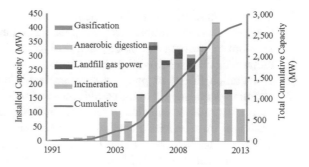

Fig. 10.9 Waste to energy
plant capacity

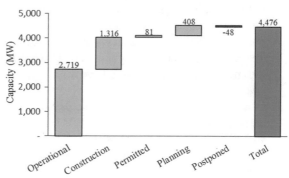

10.3.6 Challenges for Incineration

The 12th Five-year Plan of China sets a target of more than 35 % share of harmless waste disposal by incineration (The State Council 2012), which stimulated
rapid development of waste incineration power generation. Figures 10.8 and 10.9
show China's waste to energy operational plants and the total capacity status at
2013. Though incineration has great opportunity, it also faces challenges such as
variation of fuel heat value inherently brought by complex and changing waste
composition, uncertainty of MSW production in future, criticism, and skepticism
from the public on plant location and emission regulation, policy and regulation
risk on more stringent emission, less feed-in tariff and disposal subsidies.

10.4 Conclusion for the Development
of MSW Disposal Sector

Through developing a central base case scenario and exposing it to a multitude of
sensitivity analysis, the study highlights how the model can be used to inform and
provide quantitative insights to policy makers as they plan how to best maximise
the energy potential held within China's MSW sector.

(1) Under the base case scenario studied, incineration and landfill with gas collection becomes the dominant technologies, while landfill is ultimately relegated to simply absorbing the process residues from the other technology options rather than absorb MSW directly.

(2) Incineration linked to power generation is the most compelling technology option under nearly all of the sensitivities and policy scenarios. Its advantage is derived from a combination of factors: net electrical efficiency for power generation of over 16.5 %, a continuous decline rate in construction costs at over 1.5 % p.a. to 2030, and an average calorific value for waste >1800 kcal/kg. A feed-in tariff over 0.325 RMB/kWh together with future methane emission reduction and targeted landfill tax policies would strengthen its competitive position. Incineration loses its long-term economic advantage when waste disposal subsidies reduce to a certain degree or investment costs fail to improve. To meet the recent more rigid pollutant emission control standards, technology progress on incineration will have a great impact on the future of MSW disposal sector.

(3) Landfill gas to power begins to display better economics than incineration in cases where high net efficiency levels (over 46 % against a base case assumption of 28.5 %) for MSW conversion to electricity are achieved, construction costs continuously decline at over 4 %/a, or feed-in tariffs are set at levels above over 0.7 RMB/kWh.

(4) China would benefit from setting flexible tariff prices and subsidies that automatically adjust to reflect events such as technology improvements that enhance performance and reduce cost or a rise in average calorific value of MSW; a phenomenon that could result from rising national economic prosperity allied with changes in consumer behaviours. Introducing stringent regulation on kitchen waste classification, processing and disposal could realise significant benefits in terms of raising the average calorific value of MSW to enhance power generation from waste and reduce methane emissions.

(5) A tax on methane emissions, set at appropriate levels, could be an effective instrument for delivering material reductions of harmful greenhouse gas emissions from the MSW disposal sector. A rigid limit would deliver emission reductions sooner but at a higher cost as participants are forced to invest in technologies before they have fully matured. Setting a penalty cost with a cap would allow participants to incur a penalty charge until it is economically viable for them to make the necessary investments to become compliant. Delaying the point at which regulation becomes stringent could still achieve the same overall emission reduction but at a lower total accumulated cost as technologies will be more mature and both perform better and cost less.

(6) In the event that China introduced a national scheme for reducing methane emissions in MSW disposal, placing greater emphasis on feed-in tariffs (which are regulated centrally) rather than disposal subsidies (which are determined by local governments and not uniform) would allow greater control from the centre of delivery against national targets by directly influencing the economics of investments.

10.5   Update on the Policy Status of CCS

10.5.1   Introduction

China is now the world's largest greenhouse gas emitting nation. Analysis under-taken in 2014 indicates that China has now overtaken the EU in per capita car-bon emissions, producing 7.2 tonnes per person to the EU's 6.8 tonnes. At a United Nations' summit held in New York in September 2014, and ahead of a key UNFCCC meeting in Paris at the end of 2015, China pledged to take firm action on climate change, committing to make its economy much more carbon efficient by 2020 and hoping that its emissions would soon peak. Whilst the pledge may give some cause for optimism, with President Obama also declaring that the US and China have a joint responsibility to lead other nations, significant hurdles remain on the pathway to negotiate internationally a long-term successor to the Kyoto Protocol. Expectations remain that China is likely to continue to press for leniency for developing nations along with adequate financial and technological support coming from OECD nations who, it still argues, should bear the brunt of any reduction commitments. Carbon capture and storage (CCS) applications, often regarded as a critical technology solution to achieving material greenhouse gas emission reductions, are likely to lie at the heart of any resultant technology trans-fer or development collaboration agreements embedded within any future inter-national accord. Drawing primarily upon analysis and activities conducted within the BP-Tsinghua CEC collaboration, this section provides an update on the current state of greenhouse policy and CCS demonstration projects in China today.

Ultimately, the imperative for carbon capture and storage globally will be dependent on securing a long term, legally binding international agreement to address climate change. The extent to which it then becomes part of the solution, certainly in China, at this stage looks highly problematic and cannot be taken for granted. However, it is difficult to see that, with China's massive dependency on coal set to continue, it will not play a role somewhere and at some point. With this in mind, China is currently making some efforts to gain insights into its costs and operational challenges through specific pilot projects.

10.5.2   Why CCS Is Relevant

Most of the growth in the world's anthropogenic CO_2 emissions by 2050 will be due to new build coal-fired power generation in highly populated developing nations. China's recent phase of economic development has brought millions of its citizens out of poverty but at the cost of becoming both the world's largest con-sumer of energy and emitter of CO_2. The rapid expansion of China's power sec-tor from 2000 was underpinned by coal-fired generation with official statistics for 2012 indicating that coal provided 75 % of all electricity generated; this trend is unlikely to change despite the simultaneous deployment of renewable and nuclear technologies. Outside of the power sector, coal underpins the steel and cement industries and is increasingly being looked to as an alternative to oil and gas as feedstock in the downstream sector with China at the forefront of developing and deploying technologies to convert coal to gas, fuels and chemicals.

Whilst the significant underpinning of China's burgeoning economy by coal will, in the long term, contribute to the detrimental effects of climate change at a global level if left unabated, it is already creating significant environmental problems today at a local level through low-level acidifying (non-greenhouse gas) pollutants. Whilst national priorities like economic growth and energy security continue to shape China's development model, pollution reduction has become a priority for the government to address. A growing sense of urgency to reduce local pollution and improve quality of life in cities is driving China towards adopting a more sustainable approach to development and is likely to receive greater attention than implementing policies to address greenhouse gas emissions.

That said, China has already initiated pilot schemes to experiment with carbon reduction market based mechanisms and is sponsoring pilot demonstration projects for carbon capture, utilisation and storage (CCUS). Question marks remain over the likelihood of whether these pilots will be allowed to mature and scale up and, if so, at what pace? From a technology perspective, CCS is not the only carbon reduction solution available to China, for example renewables offer alternative solutions. The extent to which it becomes a compelling option will depend on the scale and timing of any reduction commitment that China may make on the international stage and to the extent to which CCS technologies can be made commercially viable relative to other options, plus the scale, viability and availability of storage facilities (Fig. 10.10).

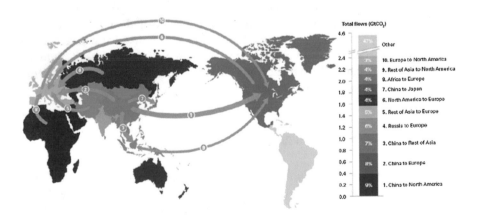

Note: Rest of Asia excludes China, Japan and India.
Data includesflow of Scope 1–3 (direct, indirect and upstream) emissions arising in region of export that are embodied in trade flows to the rgion of import.
Refer to Annex [B] for region definitions.
Source: Carbon Trust Analysis: CICERO/SEI/CMU GTAP7 EEBT (2004) Model.

Fig. 10.10 Top 10 largest inter-regional flows of embodied CO_2 emissions

10.5.3 Policy Status

In its Twelfth Five-year Plan, China set an ambition for CO_2 reduction on the basis of intensity of CO_2 against GDP for 2015 and 2020, respectively. In addition, measures to reduce energy intensity per GDP and further diversify the energy mix by increasing cleaner forms of power fuels e.g. renewables, nuclear and natural gas, are in place.

Dealing with local air pollution has become the priority item on the government's agenda. A growing sense of urgency to reduce pollution and improve quality of life in cities is driving China towards adopting a more sustainable approach to development. Within the broader pollution debate, air quality, and more specifically PM 2.5 (particulate matter of 2.5 micrometres in diameter or less), has evolved from what was originally an industry-specific challenge to become a focal point of citizens' concerns that is now difficult for the government to ignore. Addressing low-level acidifying emissions sulphur dioxide (SO_2) and oxides of nitrogen (NOx) has become a greater priority for China than tackling greenhouse gas emissions.

China's 2013 Action Plan for Air Pollution Prevention and Control (the "Action Plan") is the nation's most recent and highly visible campaign for sustainable development. Released in September 2013, the Action Plan includes a list of ten measures to reduce air pollution in major regions across China by 2017. High-level goals include capping the share of coal in China's energy mix at 65 % and reducing PM 2.5 emissions by 25 % in the Beijing, Hebei and Tianjin region (the "Jing-Jin-Ji" area), 20 % in the Yangtze River Delta and 15 % in the Pearl River Delta, based on 2012 levels. Although the coastal province Shandong and the coal-rich regions of Inner Mongolia and Shanxi are included, they are subject to less stringent targets.

10.5.4 Carbon Trading Pilots

In parallel to the "Action Plan", pilot emission trading schemes are underway in seven parts of China (Beijing, Shanghai, Tianjin, Chongqing, Shenzhen—the five cities—and Guangdong and Hubei provinces) with the potential for sectoral pilots to also be initiated under the 12th Five-Year Plan. These are designed to test a variety of individual elements and possible policy instruments rather than attempt to establish and scale up a fully integrated national approach. The success or failure of those experiments will, to a large extent, determine the future of carbon markets' development in China and having far reaching implications for carbon market development in China, not only in terms of government confidence and willingness, but also in terms of trust in a national carbon trading market. If successful, they should hopefully reveal the conditions that China would need to create to in order to incentivise CCS deployment. In terms of mechanics, each pilot has been required to submit proposals explaining how carbon emission targets will be allocated, establish a dedicated fund to support the carbon trade market and present detailed implementation plans for approval by the State Council (by the end of 2012). With five schemes launched in 2013, the goal is to have all seven trading carbon at a regional level by the end of 2014 and at a national level by 2016. Local governments have been able to decide upon the means of capping and selecting capped sectors themselves. By August 2014, Beijing, Shanghai, Guangdong, Shenzhen, Tianjin, and Hubei had completed the first compliance audit after regulated emitters surrendered their allowances. Results for Shanghai, Guangdong, Shenzhen, and Tianjin indicated an average rate of compliance of 98.85 %. Chongqing will complete its first compliance year in May 2015.

10.6 Technology and Demonstration Projects

10.6.1 Importance of CCS as a Technology for China

Given coal's continual dominance of China's long-term energy outlook, a future scenario involving joint international efforts to reduce greenhouse gas emissions means that CCS as a major contributor to compliance in China cannot be overlooked. However, widespread deployment of CCS, even outside of China, is not given and faces many hurdles. One leading challenge is that of efficiency with decarbonisation of coal (and other fossil energy sources) in power generation via CCS resulting in an efficiency penalty of 15–30 %. Consequently, more primary fuel inputs are needed to generate the same amount of electricity; significantly increasing the cost of electricity produced from CCS-enabled plant and countering any energy conservation policy initiatives.

Preserving economic growth and its position in global markets could become a compelling source of value for CCS in China. In the event that binding international carbon-emission constraint agreements requiring embedded CO_2 emissions to be priced into goods traded between nations came into place, this would have a negative impact on Chinese exports. It is estimated that, in 2004, ~23 % of China's emissions were 'exported' as commodity production, especially manufactured goods exported to the US.

10.6.2 Technology Status in China

With no sign of abatement, China's continued heavy reliance on coal to fuel its burgeoning fleet of relatively efficient and relatively cheap to build coal-fired power stations, coupled with the potential risks associated with producing half of the world's CO_2 emissions would seem to offer a compelling case for CCS. In addition, given its success in making advances in coal power and coal conversion technologies, CCS could similarly hold distinctive advantages as a technology for China to gain leadership in. These nuances are recognised with the Ministry of Science and Technology (MOST) providing scientific research and policy support and CCS having been integrated into a number of China's national technology programmes since 2005. These include the 973 Programme (2006) and CCS elements of the 863 Programme (2008) funded by MOST to support research across capture, storage, EOR and verification as well as coal gasification demonstrations for power and coal to chemicals.

There are 10 demonstration projects involving major companies Huaneng, Shenhua, Sinopec, Petrochina, and Shaanxi Yanchang already in operation, predominantly addressing individual CCUS elements. The largest demonstration planned is in Tianjin Province by GreenGen; a consortium made up of the leading local power and coal players Huaneng, Datang, Huadian, Guodian, China Power Investment, Shenhua, State Development & Investment Co., China Coal and one international player, Peabody (the world's largest private sector coal company). It centres on Integrated Gasification Combined Cycle (IGCC) power generation technology with coal as its feedstock. The power plant was put into operation in Nov 2012 which marked the completion of GreenGen Stage I with a generation capacity of 250 MW including a 2000 t/d dry-coal powder gasifier. Stage II is set to see a CCS demonstration of up to 100,000 t/a CO_2 by 2015 and Stage III to see building of a 400 GW IGCC and associated CCS.

Huaneng's 100,000 tonne per annum IGCC based pre-combustion CO_2 capture pilot is due to commence in 2015 though no associated utilisation/storage project has yet been identified. However, IGCC technology, once touted as the saviour of coal power and the future of clean coal generation, has seen its prospects deteriorate by soaring costs and technological challenges. Therefore, it is worth noting that Huaneng also has a 100,000 tonnes/year CO_2 *post-combustion* capture system demonstration project at its Shanghai Shidongkou No. 2 Power Plant. This has been in operation for approximately 5 years but, as yet, with no associated storage demonstration.

However, despite developments in carbon capture technologies, these alone will not address emission issues. In addition, some carbon capture and utilisation (CCU) pathways into industrial applications simply delay the release of CO_2 since the storage is only temporary (days/months). Utilisation for enhanced oil and gas recovery (EOR and EGR) would offer larger scale and higher value propositions from additional oil/gas production.

Of the seven types of storage technology options identified, the most feasible with large-scale options for China will be deep saline formations unrelated to hydrocarbon production (absent sufficiently large depleted gas and oil fields or coal seams). Characterisation of the potential storage capacity available from Chinas' geology has yet to be completed, internationally compared and verified.

However, multiple smaller scale storage of near pure stream CO_2 from polygeneration of power, synthesis gas (syngas) and products from coal could be a useful bridge to a storage-focused solution while bringing together multiple stakeholders and existing research knowledge.

At an international CCS road-mapping workshop hosted by the CEC, local and international attendees shared different views on how China may pursue CCUS over the coming decades. A consistent view was recognition that, with the United States largely decarbonising through improved energy efficiency and shifting to shale gas, the eyes of the world would fall increasingly on China and its coal sector. It was also felt that while the global economic issues of recent years had diverted Western governments' priorities away from the climate debate, this was seen only as a temporary respite.

Nevertheless, even with an eventual return of global consensus for action, the uncertainty would still remain around the unproven economic and practical viability of CCS. Hence this would still require a well-planned and well-managed government policy to deliver successful demonstration programmes and associated fiscal incentives and regulatory structures to create the right environment for CCS deployment. Local experts also expressed the view that China will need to fully address the CCS question in order to protect its ongoing ability to consume coal to sustain long-term economic growth. The following table summarises some of the key perspectives.

China experts	International experts

• Value for China is ongoing ability to draw on domestic coal to power economic growth
• Demonstration requires $30/t carbon price by 2030, building on 3 × 1 Mt storage demos by 2025
• Success factors:
  – Technology transfer
  – Ongoing climate concerns
  – Carbon price increase/trading
  – Govt. support
  – Economic stability
  – America takes the lead
	• China can take lead by setting global standards in reservoir characterisation/integrity standards
	• Demonstration chain at $50/t storage costs by 2030 with 75 Gt of proven storage
	• Success factors:
	  – Storage resource mapping
	  – Energy systems modelling
	  – International knowledge sharing/collaboration
	  – Endorsement by Government as official plan
	  – Academia convenes and facilitates dialogue

The common themes are the importance of the roles of government, of international cooperation and the time horizon for significant implementation being 2030.

10.7 Conclusion and Suggestion

10.7.1 Future Hurdles

A lack of significant progress in the CCUS programme globally has essentially created a very negative perception, sending mixed messages to policy makers. This has no doubt been exacerbated by apparent changes in OECD priorities and perceived attitudes to climate change prior to the successful Paris COP-21 meeting in December 2015. Consequently, the activities in this space in China have appeared to lack cohesion and direction with demonstration pilots increasingly being undertaken without obvious reference to an integrated framework or guiding structure that encompasses all elements of the chain.

In 2011, a China CCUS Roadmap, a national level study covering the whole CCUS technology and policy landscape, was developed to help set a clear national direction. While Government support for research and development has been provided via the 072 Fund (with PetroChina in particular involved) significant uncertainty remains on China's storage potential. With a clear lack of geological data available, this remains a crucial missing element to the purpose and success of any CO_2 sequestration programme. Until this is resolved, it is likely to undermine confidence in moving forward. In addition, given China's relatively limited hydrocarbon resource base today, options for valuable EOR and EGR are unlikely to feature highly.

10.7.2 Future Opportunities

Developing a more coordinated framework for these activities will help China demonstrate global leadership and in a manner that encompasses all of the benefits of international cooperation. Leadership does not necessarily mean going first or going it alone, rather recognising that China has the potential to lead in terms of engaging other nations on the benefits of more integrated global cooperation on specific elements that will deliver large-scale full-chain CCS demonstration, verification and regulation. There is a mutual benefit in all parties understanding and gaining line of sight of the cost of CCS. This would have a significant bearing on the commitments that nations might be willing to sign up to as part of any future climate change reduction agreement.

Through international cooperation and taking leadership, China could place itself at the forefront of discovering the critical cost points for CCS. This will be vital in allowing Nations to evaluate where CCS sits, both in terms of cost and CO_2 mitigation relative to other options available to them. This could also help move the debate away from "who pays now" into a longer term plan where demonstration investments have clear rationale, validity and mutual value at each step. Given China's proven track record of manufacturing and cost innovation, helping OECD nations understand where CCS might sit within their national portfolios of carbon abatement options could potentially create new markets for China to export CCS know how.

10.7.3 Conclusion for the Role of CCS

Similar to the attempts to secure a legally binding international agreement on climate change, delivering CCS will not be straightforward and face many hurdles. It requires a strong and sustained policy impetus. As a solution, it does not consist of a single technology and will inherently involve large, complex, capital-intensive, and expensive projects. As a process, it spans many disciplines and integrating technical and operational expertise from the coal, power, chemical, and oil and gas sectors, which will be a demanding yet necessary requisite. In terms of storage, limited options for supporting enhanced recovery in oil and gas leaves saline aquifers as the leading option but the exploration, appraisal and development of suitable sites holds no value proposition unless carbon is priced high, and certainly at levels above which many other non-CCS options would be commercially viable.

Then we have to consider the competition that CCS faces. Many competing solutions such as wind, solar, and nuclear are available; these are either reaching maturity at or close to grid parity with incumbent coal in China. In nuclear, China has recently accelerated its programme to develop and commercialise thorium based technologies which is abundant as a resource and produces significantly lower waste toxicity than uranium. These are all technology areas where China has proven track records in manufacturing and deployment.

Ultimately, the imperative for CCS globally is wholly dependent on securing a long term, legally binding international agreement to addressing climate change. However, it is difficult to see that whether it will play a role somewhere or at some point.

References

National Bureau of Statistics of the People's Republic of China (2014) Chinese statistic yearbook 2014. China Statistics Press, Beijing

The twelfth five-year national urban household solid waste disposal facilities construction plan (2012) The state council, the People's Republic of China. 29th Apr 2012

Yuanwei W, Yali X, Jiongyu Y, Weidou N (2013) Prediction of municipal solid waste generation in China by multiple linear regression method. Int J Comput Appl 35(3):1–5

Chapter 11
Conclusions

China has become an economic superpower over the past 20 years and will continue to grow substantially for the foreseeable future. All nations require energy to fuel economic growth. It is increasingly obvious that China needs to find a more sustainable path to power this economic growth.

This publication has attempted to describe the scale and complexity of China's energy system. We have created and used Sankey diagrams to visualise the energy flows across the economy. We have also looked to visualise and categorise the differences between China's provinces; China's provinces extend over a vast geography and are at different stages of economic development, regional planning. Inter-regional optimisation will be essential for planning future infrastructure that will allow China as a whole to optimise and avoid overcapacity and inefficiency.

Consistent with this analysis, we have introduced our energy systems modelling approach and demonstrated its application to the power sector. Our power model can help to optimise the future design of the regional power systems and reveals how different future investment choices emerge as the level of detail is increased, such as regional, seasonal and diurnal demand factors are considered. We have also started to look at options for mitigating carbon and the major role that renewables and natural gas could play as well as carbon pricing. Our waste to energy analysis has demonstrated how energy systems modelling can also contribute and give insights into how policy can be designed to deliver optimal outcomes for China.

Our work will continue to improve the power model, including enhanced representation of emission reduction technologies. Planning a more sustainable future is now critically important for China.

© Springer Science+Business Media Singapore 2016
Z. Li et al., *Informing Choices for Meeting China's Energy Challenges*,
DOI 10.1007/978-981-10-2353-8_11

Glossary

BIGCC Biomass Integrated Gasification Combined Cycle

CBM Coal-bed methane

CEPP China electric power press

CFB Circulating fluidized bed

CHP Combined heat and power

CMM Coal mine methane

CNG Compressed natural gas

COE Cost of electricity

CPECC China Power Engineering Construction Corp

deNO$_x$ Denitration

gce Gram of coal equivalent

GE General Electric Co

HD Hydropower

HRSG Heat recovery and steam generator

HTGR High temperature gas-cooled reactor

IEA International energy agency

IGCC Integrated gasification combined cycle

IRR Internal rate of return

ITER International thermonuclear experimental reactor

LHV Low heating value

© Springer Science+Business Media Singapore 2016
Z. Li et al., *Informing Choices for Meeting China's Energy Challenges*,
DOI 10.1007/978-981-10-2353-8

LNG Liquefied natural gas

MHI Mitsubishi Heavy Industry

MLR Ministry of Land and Resources

NBS National Bureau of Statics

NDRC National Development and Reform Commission

NGCC Natural Gas Combined Cycle

NO$_x$ NO and NO$_2$

OID Oil import dependency

PC Pulverized coal

PM Particulate matter

PV Solar photovoltaic

PWR Pressurized water reactor

SCR Selective catalytic reduction

SOE State-owned enterprises

SPC Supercritical PC

SPR Strategic petroleum reserves

sub-PC Subcritical PC

tce Tonnes of coal equivalent

USPC Ultrasupercritical PC

VAT Value-added tax

Printed in the United States
By Bookmasters